Technology, Culture, and Public Policy

Dr Kalu N. Kalu gives an elegant account on how Finland transformed itself from an agricultural society to one of the world's most inventive countries in just a century. Education, equal rights and willingness to re-invent itself are in the heart of this story.

– *Jouni Mölsä, Ministry for Foreign Affairs of Finland, Finland*

In a relatively short time, Finland has transformed a society of approximately 5.3 million people into one of the most educated and technologically sophisticated in the world, while maintaining relative political stability and an enviable quality of life among its people. In all comparative measures of international achievement, Finland ranks among the world's most literate and wealthiest countries. How did Finland do it, and what can other countries learn from the Finnish example?

This book presents an energized and informative look at Finland's cultural and developmental history, its political evolution as a state, the origins of its technology and innovation policy, and its developments in health care, education, and sustainable economic development. Utilizing both qualitative and quantitative approaches, author Kalu N. Kalu incorporates rarely-seen archival data alongside analysis of original research surveys to offer an incisive and analytical account of virtually all aspects of Finnish life. Demonstrating how civic attitudes have evolved over time mediated by contemporary pressures, *Technology, Culture, and Public Policy* ultimately transcends an examination of Finland's own successes and challenges, considering what lessons other countries might apply to their own intricate national contexts.

Kalu N. Kalu is Distinguished Research Professor of Political Science & National Security Policy at Auburn University Montgomery, USA and Docent Professor at the University of Tampere, Finland.

Technology, Culture, and Public Policy

Critical Lessons from Finland

Kalu N. Kalu

Routledge
Taylor & Francis Group
NEW YORK AND LONDON

First published 2017
by Routledge
605 Third Avenue, New York, NY 10017

and by Routledge
2 Park Square, Milton Park, Abingdon, Oxon OX14 4RN

Routledge is an imprint of the Taylor & Francis Group, an informa business

Library of Congress Cataloging in Publication Data
Names: Kalu, Kalu Ndukwe, author.
Title: Technology, culture, and public policy: critical lessons from Finland / by Kalu N. Kalu.
Description: New York: Routledge, 2016. | Includes bibliographical references and index.
Identifiers: LCCN 2016004522 | ISBN 9781482223446 (cloth: alk. paper) | ISBN 9781315563428 (ebook)
Subjects: LCSH: Finland—History—1981- | Technological innovations—Government policy—Finland. | Finland—Cultural policy. | Finland—Economic policy. | Finland—Social policy. | Finland—Politics and government—1981-
Classification: LCC DL1140.6 .K35 2016 | DDC 320.6094897—dc23
LC record available at http://lccn.loc.gov/2016004522

ISBN: 978-1-4822-2344-6 (hbk)
ISBN: 978-1-315-56342-8 (ebk)

Typeset in Times New Roman
by codeMantra

For
Elesie Ukanwa Ndukwe Kalu

Contents

List of Figures and Tables

Figures

Tables

Map of Finland

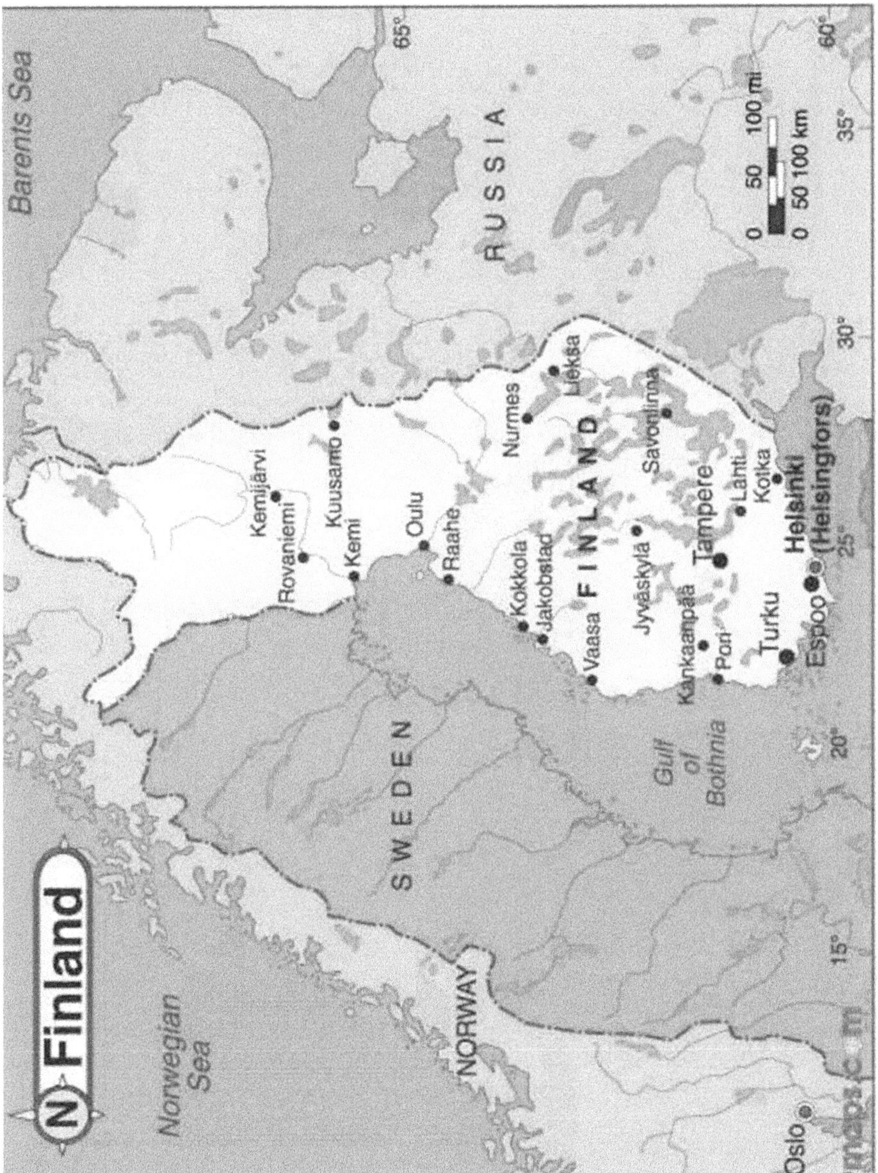

Preface

Finland ranks at the top of several measures of social and economic development and has exceled in the areas of education, quality of health care, technology, and innovation. Yet, the country has maintained relative political stability in the way its Nordic welfare state has operated over the years. As a technology-intensive society, we can, at best, expect to get a better understanding of how technology induces and sustains economic development and innovation, bridges ideological or cultural differences, and may influence citizens' assessment of the effectiveness of political and administrative institutions in addressing key social problems that fall under their programmatic jurisdictions.

This choice reflects my long-standing interest and work on the role of information technology in reshaping organizations and society and redefining functional roles as well as the cultural foundations upon which social and political life has generally existed. What type of cultural and organizational adaptations has technology brought to Finland, and how has it advanced its political, social, and administrative development? Under the current situation, how would Finland evolve in the future and in what areas? What could other countries in general learn from Finland? These are some of the major questions that justify a study of the country.

Chapter 1 traces the cultural foundations of Finland, its political and economic history as well as major historical events that have shaped its evolution as an independent sovereign state. It discusses how the seeming disadvantages of geography and harsh weather were turned into a success story admired by the rest of the world. It also illustrates early efforts at industrialization beyond the long period of political incubation under Swedish rule, reliance on forest and wood products, its search for identity, the rise of literary and intellectual ferment, and the consequent drive toward independence from the Soviet Union after having fought two earlier wars with it.

Chapter 2 provides an extended discussion of the evolution of Finland's political history after more than 700 years under Swedish rule and its consequent quasi-autonomous status as a Grand Duchy of the Soviet Union. It traces the inspiration and history of the independence movement, the experience of the civil war between the Bolshevik (Reds) and the conservative (Whites) immediately following independence from the Soviet Union in 1917. There are three important characteristics that have continued to shape Finland's political culture—equality, egalitarianism, and consensual politics, particularly in the area of legislative or parliamentary policy making. The Parliament of Finland is discussed as a learning organization, with multiple committee specializations and membership, that cuts across partisan differences.

Chapter 3 discusses the emergence of Finland from a rural-agricultural economy to an advanced technology-driven economy within a short span of years. It discusses the role of state dirigisme in laying the early foundation and support that saw increased cooperation between the public and private sectors, between educational and research institutions, and industrial manufacturing that saw scientific knowledge transformed into value-added production for the export market. It delineates the origins and foundation of Finnish technology policy which began in the early 1960s and 1970s, but saw its most profound manifestation following the economic recession and decline in export trade caused partly by the fall of the Soviet Union in 1991. Technology policy and innovation-driven macroeconomic policy guided by the *dirigiste* state in partnership with the private sector is discussed as the backbone of Finland's technology success story.

Chapter 4 goes in greater detail to illuminate the processes by which technological development became a central factor in Finland's economic growth, the major institutional actors in terms of universities, research institutes such as the Academy of Finland, Tekes, and the Finnish Innovation Fund (SITRA). It also discusses the objective and principles in the development of innovation clusters and networks—as a means of building a sustainable value chain between and across technology-driven regional innovation networks.

Chapter 5 lays out in stark detail the overall principle, design, and structure of Finnish education policy and why it has become celebrated around the world. On all competitive measures of educational achievement, particularly in math and the sciences, Finnish students have proven to score above the rest. How did Finland do it? What can the rest of the world learn from Finland? Finland has universal free education at practically all levels, but also emphasizes catching non-performing students very early in their educational program so that appropriate remedial measures can be taken before their situation gets worse. By laying greater emphasis on basic education at the lower primary levels, the educational system goes to great lengths to prepare students for the learning project after they take their matriculation examination in readiness for either upper secondary education or vocational training.

Chapter 6 discuses Finland's health care system, which is based on a Nordic-style welfare system covering universal health insurance, primary and occupational health, and an emphasis on nutritional and healthy-living practices. Its structure is built around municipal and district hospitals and a capitated payment system that covers essentially the whole life cycle of care. While employers are bound by law to ensure health and safety for workers and also contribute to health care funding, state subsidies to municipal social welfare and health care expenditures are determined by population, age structure, and morbidity in the municipality plus a number of other unique factors such as income levels. Finland's health care system is ranked among the best in the world even though it offers universal coverage for everyone.

Chapters 7 and 8 focus on how technology can drive culture and value change thereby casting specific influences on the policy assessment process both in terms of a general civic trust of the national government and in the efficacy of specific programs, particularly in education and health policies. They address the design and dissemination of several survey questionnaires, collection of data, transformation and analysis. By utilizing more sensitive methodological approaches and reliability scales to further advance several of the hypotheses suggested, it can be surmised that while technology is a bridging mechanism, its relationship to culture may be reciprocal.

Chapter 9 addresses issues and challenges of a potential future society and how adaptation to its uncertain challenges may be mitigated by the application of novel and even exotic technological breakthroughs. But then, what if that which was hitherto thought impossible becomes possible? How would mankind adapt to a future world that is unpredictable, one in which global catastrophes and *black swans*—unlikely but unforeseen events that change the world—occur unexpectedly? The Committee for the Future (created by the Parliament of Finland) has been working arduously to seek answers to these and other very vexing global challenges. The chapter also discusses future and emerging technologies and the role of Finland, which has remained at the forefront in the development of what is known as *radical technologies*.

Chapter 10 presents a nostalgic commentary on essentially all critical aspects of Finnish life—from the arts, music, sports, architecture, literature, fashion, and color, to the more enduring aspects of culture such as leisure, communication, and the place of *sauna* in Finnish tradition. The Finns take great care infusing meaning and context in their arts, design, and sculpture, and this quality is also very much reflected in their architectural and building designs, in their clothing and apparel, and in the materials they use in the transformation process for these works. Collectively, these can help point to a central theme, one that is both abstract and practical; but also one that, in essence, could be said to resonate in its own peculiar narrative, a working portrait of an unvarnished Finnish character—all the way.

Kalu Ndukwe Kalu
Auburn University Montgomery, USA
Docent, University of Tampere, Finland
May 1, 2016

Acknowledgments

At the end of my first visit to Finland in 2009, which lasted for about four days, there was a certain feeling of unyielding admiration and amazement that stuck with me for a period of months. It was a sentiment born out of curiosity both for country and the people in terms of the advanced level of development, quality of life, and relative political stability—all achieved within a relatively short period of time. I then decided that there was a lot more about this country that needed to be further explored, particularly in the areas of comparative development and institutions, culture, technology, and innovation policies. Despite some trying and challenging moments during the period of this investigation, Providence strengthened my resolve to soldier on. The project is now a mission accomplished, thanks to the financial support I received as a result of the Fulbright Scholar award. It was a project that drew a lot of commitment and above all encouragement and support from many individuals and agencies both within and outside the United States. Justice therefore demands that I acknowledge them.

I am very grateful to the Fulbright Scholar Program sponsored by the U.S. Department of State and administered by the Council for International Exchange of Scholars; the institutional support of the Fulbright Center, Finland; Auburn University Montgomery (USA); and the School of Management/Politics of the University of Tampere, Finland. There were individuals who played indispensable roles that contributed severally to bringing this work to fruition. Among these are Professor Markku Sotarauta, erstwhile Dean of the School of Management, University of Tampere, Finland; Terhi Mölsä, the tireless Executive Director of the Fulbright Center, Finland; Aki Kalliomäki and Sonja Kuosmanen, Program Director and Coordinator, respectively, both of the Fulbright Center, Finland; Tapio Raunio and Tapani Turkka, Professors at the School of Management, University of Tampere; Sari Saastamoinen, Head of Administrative Staff, School of Management, University of Tampere; and the indomitable Dr. Paula Tiihonen, Committee Secretary and Counsellor to the Committee for the Future, The Parliament of Finland, Helsinki.

I thank the following individuals who provided critical information and assistance in various ways: Jouni Mölsä, Director General, Department for Communications and Culture, Ministry for Foreign Affairs, Finland; Timo Turja, Chief Communications Officer, Library of the Parliament of Finland; Jorma Kauppinen,

Director General Education, Finnish National Board of Education; Aki Tornberg, Counsellor of Education, Ministry of Education and Culture, Finland; Tapio Antilla, Executive Vice President, SITRA– The Finnish Innovation Fund; Markku Pekurinen, Director and Research Professor, National Institute for Health and Welfare, Finland; Pertti Hölttä, Chief Specialist, Finnish Communications Regulatory Authority (FICORA); Joonas Orkola, Communications Market Specialist, FICORA; Petteri Rautaporras, Economist, The Federation of Finnish Technology Industries; and Aimo Minkkinen, Director, The Lenin Museum, Tampere, Finland.

My profound gratitude to Markku Mahkonen, Director of University Relations, Nokia Headquarters, Espoo, Finland; Timo Niklas-Salminen, a former soldier and industrial shoemaker with whom I had a chance meeting on October 28, 2013, at the café in the popular department store Stockman. He eventually invited me to visit his home on November 2, 2013, to meet his family. Our follow-up discussion about all aspects of Finland proved to be priceless. I also want to thank Shahidul "Litu" Islam, who unsparingly gave his time to take me around Rovaniemi in the northern Laplands, during my research trip to the Arctic Circle.

I have to acknowledge my dearest daughter Renee, for her prayers and constant phone calls to check on my welfare while I was away in Finland during the course of this project. It was very gracious of her to call to sing me a happy birthday song over the phone in Finland from miles away in the United States. I cannot thank her enough for all her encouragements and inspiring words throughout this process. In the end, I owe a sincere debt of gratitude to the Finnish people for providing the opportunity to experience and learn about their country, institutions, and contribution to global technological development and innovation. This book is dedicated to the courage and memory of my wife Elesie Ukanwa Ndukwe Kalu.

About the Author

Kalu N. Kalu is Distinguished Research Professor of Political Science & National Security Policy at Auburn University Montgomery; Docent Professor at the University of Tampere, Finland; and a Fulbright Scholar. He has been a Research Affiliate at The Whitney and Betty Macmillan Center for International and Area Studies, Yale University, and a Visiting Senior Research Fellow at the Nigerian Institute of International Affairs. He is also a recipient of two certificates in Joint Strategic Leadership and National Security Decision Making from the United States Air War College (2010-2011). Professor Kalu is widely published with articles in leading peer-reviewed journals and has a combined production of over 60 scholarly works cumulatively in terms of publications in journals, books, edited volumes and in national and international conference presentations.

His research emphasis is in the areas of comparative institutions and development, organization theory and organizational change, citizenship and administrative theory, technology-leadership-culture interface, complex adaptive systems, national security and intelligence policy, and health care politics and policies. He has been invited and participated in several international learned conferences and fellowships in various countries around the world including Canada, Germany, Italy, Norway, Finland, Switzerland, Israel, Saudi Arabia, United Arab Emirates, China, Singapore, and India. He has served on the Editorial Board of *Public Administration Review*, the premier journal in the field. He is the author of *"State Power, Autarchy, and Political Conquest in Nigerian Federalism" (Rowman & Littlefield Publishers, 2008);* and *"Citizenship: A Reality Far from Ideal"* (co-edited with Nada Kakabadse and Andrew Kakabadse, 2009).

1 Suomi

The Making of a Culture

Finland (*Suomi*) is known to have a history of determination, creativity and industriousness that has propelled the country into being a pioneer of education, modernization, inventiveness and technological prowess. It has also managed to lay an enviable foundation of sustained social and economic development, partly premised on advancements and emphasis on cutting-edge information technology culture, learning and educational systems that provide the creative energy and innovation for future growth and development. Officially the *Republic of Finland*,[1] is a Nordic country situated in the Fennoscandian region of Northern Europe. It is bordered by Sweden to the west, Norway to the north and Russia to the east, while Estonia lies to the south across the Gulf of Finland.

1.1 Introduction

An estimated 5.4 million people live in Finland, with the majority concentrated in its southern regions. The two major official languages are Finnish (90.4%) and Swedish (5.4%), although English is also relatively used equally for official purposes. The main religious affiliations are the Evangelic Lutheran Church (78%) and Orthodox Church (1%) of the population. About 19% of the population subscribe to no affiliation. In terms of area, it is the eighth largest country in Europe and the most sparsely populated country in the European Union. Politically, it is a parliamentary republic with a central government based in the capital of Helsinki, six provinces, local governments in 336 municipalities and an autonomous region, the Åland Islands. About one million residents live in the Greater Helsinki area (consisting of Helsinki, Espoo, Kauniainen and Vantaa), and a third of the country's GDP is produced there. Other larger cities include Tampere, Turku, Oulu, Jyväskylä, Lahti and Kuopio.

Finland was a relative latecomer to industrialization, remaining a largely agrarian country until the 1950s. Thereafter, economic development was rapid, such that today, with a nominal per-capita income of over $49,000 (2011 estimates), Finland is one of the world's wealthiest nations. According to some measures, Finland has the best educational system in Europe and has recently been ranked as one of the world's most peaceful and economically competitive countries. It has also been ranked as having one of the world's highest quality of life. Basic studies in the humanities as well as the sciences are equally appreciated as part of the educational system, in such a way that 'hard' and 'soft' values are in harmony and diverse cultural activities, supported from public funds, counterbalance science as an important part of a stable constitutional welfare state. "International studies now rate Finland as one of the world's most competitive and innovative countries, and foreign visitors come to learn of its ethical corporate culture."[2]

I visited the country in 2009 while attending the 5th Specialized International Conference of the *International Institute of Administrative Sciences* in Helsinki and was quite impressed by a social system in transformation anchored on a proactive partnership between the government, its institutions, and the society at large. Compared to many other countries of Europe, Finland's modern welfare state, its high standard of education, equality promotion and emphasis in the application of information technology to education and the delivery of public services makes it a study in contrast. While it spends about 11.7% of its GDP on health and about 6% on education, it has maintained a 100% literacy rate with a per capita income among the highest in Western Europe. Finland's economy is anchored on a highly industrialized free-market system and is competitively strong in the engineering, telecommunications, and electronics industries. What makes Finland unique is that it has over a relatively short time transformed a society of approximately 5.4 million people into one of the most technologically sophisticated in the world (knowledge society) and has at the same time maintained relative social stability and an enviable quality of life among its people.

Finland is ranked among the world's most inventive countries, and much of this can be attributed to the close partnership between businesses and universities. Universities and research centers have good resources for high-quality teaching and research, and the educational opportunities available to students remain among the best in the world. How did Finland do it? How has it utilized information technology to sustain a stable and progressive welfare state? How does technology transcend existing cultural practices, and in what ways do these cultural practices influence the way citizens respond to or evaluate policies and programs from political and administrative institutions? What can other countries learn from the Finnish experience? These are some of the reasons a further study about various aspects of Finland's development is necessary.

Why Culture?

As argued in earlier studies,[3] cultures emerge in all human societies from people's struggles to manage uncertainties and create some degree of order in social life. As a way of helping to manage these uncertainties and ambiguities, cultures provide people with accepted ways of expressing and affirming their beliefs, values and norms—and provide them with cultural repertoires or forms that members of a community use to express the substance as well as make sense of their respective worlds.[4] Cultural forms imbue action with meaning; they enable people to communicate and celebrate their ideologies in many different ways. Cultures cannot be produced by individuals acting alone; rather they originate and endure as individual members of a community interact with one another. Hence, belonging to culture involves believing what others believe and doing as they do.

In turbulent times, culture helps a people to manage their anxieties and disagreements and to channel their individual emotions and reservations in such a way that it does not fundamentally undermine the socially established mechanism for achieving deliberative consensus and mutual co-existence.

Though it may seem inherently symbolic, a society's cultural disposition transcends its own history, context and experiences. Although a body of literature has emerged that underscores the relationship of IT and culture, the larger premise of this project is that culture at the national or institutional level can influence the way implementation and use of information technology unfolds and vice versa. While culture at the national, organizational or subunit level may exert a subtle and yet powerful influence on people and organizations, information flows and information technologies are often closely intertwined with culture.[5] Culture theory has been used in studies that seek to explain social behaviors and outcomes in organizational settings,[6] administrative practices, technology transfer practices, the development and use of ICTs, and the potential for cultural transformation through technology use.

Hence, to the extent that information technology use may be a reflection of social and economic development, it is still subject to the various cultural interpretations of those using it. These interpretations may influence changes in civic attitudes and/or dispositions toward political institutions and the social values (social capital) attached to the utility and effectiveness of their programmatic responsibilities and initiatives. In this regard, Finland serves as a model case. Among the Finns, cultural expressions are at the center of everyday lives, and they find fulfillment in each mundane and ordinary undertaking. It is a life in which almost everything is nuanced or rather "reasoned," even in seating arrangements: There must be a balance on both sides instead of everyone or most of the people sitting on one side. Equally, the Finns are very proud of their language and give very little concern to speaking it in meetings in the presence of foreigners, even when it is obvious to them that the foreigner does not speak or understand Finnish. It is this self-awareness and an almost ubiquitous validation of Finnish identity through language that is at the core of being a Finn. There is the strong belief that when you are in an environment you have to reinvent yourself to adjust to the prevailing circumstances and challenges of that environment.

The Finns are very practical and particularly thoughtful in their approach to things and human events. There are three broad elements that define Finnish culture. First is a marriage of grand idealism, futurism and a pioneering tradition rooted in a mindset that continuously seeks practical solutions to all of mankind's enduring problems. Second, it is a culture built on a national vision for a larger global future characterized by rapid renewal through social innovation. One such innovation could be tied to the post-civil war attitude of the Finns as they decided to rebuild the nation on the principle of equality. A central question that is often asked about the Finnish vision is: What do we need in order to resolve the many difficult issues facing the world? And from this would emerge a not-so surprising answer: experience (in solving problems), ingenuity (in finding inventive solutions to new problems) and courage (in taking on problems that seem impossible to solve). Inventiveness and a do-it-yourself spirit offer a pathway to unique and unusual solutions.

Changes in the Production Structure

Finland's agricultural productivity accounted for 62% of its GDP in 1860 but declined to about 30% by 1950, while the share of total industry started at .08% in 1860 but rose to 30% in 1950. As a result of economies of scale and market competition, total industrial production as a percentage of GDP remained at 20% in 2010, while the share of agricultural productivity declined to approximately .04% of GDP.[7] What can be observed is that in the intermediate period of Finland's industrial development, the country did not demonstrate obvious manufacturing capacity that could compete comparatively in the international market. Much of that changed after World War II, due partly to the increased demand for wood and paper products and the requisite need to develop new paper machines as well as the forestry economy.[8] Today, industrial productivity is about 30% of GDP; services at 67% of GDP; agriculture and exports account for .03% and 39%, respectively; and an approximate GDP per capita of Euro 35,559.

Historically, although Finland's industrial development relied heavily on the forestry industry, it imported most of its paper machines up until the 1950s. "The first machines were exported from Finland during the period that it was paying war reparations to the Soviet Union, but by the 1990s Finland was the world leader in paper machine industry and paper manufacturing technology."[9] By instituting a series of national policy development objectives such as technological education and research, investment in machine building and innovation, developments in paper technology, research and product development tied to the paper machine industry and a strong political will to see it through, Finland was able to turn the suffocating political pressure for reparations into a potent economic and technological advantage.

As pointed out by Nykanen and Paulapuro,[10] the increased use of paper in Finland is directly linked to university teaching, and it was only when the University of Turku started its own printing press in 1641 that the amount of literature grew east of the Gulf of Bothnia. Most prominent in this was when Bishop Johannes Gezelius Senior started his own printing press in 1688 to spread Protestant literature to the people. The oldest Finnish paper mill was started in Finland during the Swedish rule. Then the German paper expert Oberhar built a paper mill on Thornasbole Estate in the parish of Pohja in 1667. Ownership of the mill was immediately transferred to Bishop Gezekius Senior, and the operation of the mill was incorporated into the production of his printing press. Furthermore, electrical engineering arrived in Finland at a very early stage once Thomas Edison created the opportunity for the extensive utilization of electricity in manufacturing. In 1882, the Finlayson factory in Tampere was fitted with electrical lighting a mere two years after the electric light bulb had been introduced to the world.[11] It would therefore be very difficult to separate the rising need for paper products and an early national emphasis on education, research and innovation as the building blocks of Finland's contemporary economic and technological prowess.

Conflict and Adversity: Shaping the Cultural Mindset

While cultures create continuity and persist across generations of members, they are not static, but rather dynamic. Cultures continually change and adapt, more or less as a result of changes in the larger political and social environment of which they are a part.[12] The cumulative effects of the Finnish people's historical and political experiences have played a central role in shaping the nationalistic mindset of the Finns as well as the extant political culture. It is an experience, though regrettable in certain respects, that has worked well to reinforce the Finnish attitude and its place in the community of nations. At critical historical junctures, the resultant effects of these experiences have turned out to be quite epic and enduring in their manifestations. But then, it would be safe to ask a question that has been asked many times before: Who are the Finns and from whence do they originate?

According to historical highlights provided by the United States Department of State Background Notes,[13] the origins of the Finnish people are still a matter of conjecture, although many scholars argue that their original home was in what is now central Siberia. The Finns arrived in their present territory thousands of years ago, pushing the indigenous Lapps into the more remote northern regions. Finnish and Lappish—the language of Finland's small Lapp minority—both are Finno-Ugric languages and are in the Uralic rather than the Indo-European family. Finland's nearly 700-year association with the Kingdom of Sweden began in 1154 with the introduction of Christianity by Sweden's King Eric. During the ensuing centuries, Finland played an important role in the political life of the Swedish-Finnish realm, and Finnish soldiers often predominated in Swedish armies. Finns also formed a significant proportion of the first "Swedish" settlers in seventeenth-century America.

Following Finland's incorporation into Sweden in the twelfth century (twelfth to nineteenth century), Swedish became the dominant language, although Finnish recovered its predominance after a nineteenth-century resurgence of Finnish nationalism. A publication in 1835 by Elias Lönnrot of the Finnish national epic, *The Kalevala*—a collection of traditional oral folklore, mythology and legends—first stirred the nationalism that later led to Finland's independence from Russia. In 1809, Finland (along with Sweden) was conquered by the armies of Czar Alexander I and thereafter remained an autonomous Grand Duchy connected with the Russian Empire until the end of 1917. On December 6, 1917, shortly after the Bolshevik Revolution in Russia, Finland declared its independence. In 1918, the country experienced a brief but bitter civil war that colored domestic politics for many years, a civil war in which the pro-Bolshevik "Reds" were defeated by the pro-conservative "Whites" with support from the German Empire. After a brief attempt to establish a monarchy in the country, Finland became a republic.

Finland's experience of World War II involved three separate conflicts: the Winter War (1939–1940) and Continuation War (1941–1944) against the Soviet

Union and the Lapland War (1944–1945) against Germany as the Nazi Army withdrew its forces from northern Finland. Following the end of the war, Finland joined the United Nations in 1955, the Organization for Economic Co-operation and Development (OECD) in 1969, the European Union in 1995 and the Eurozone at its inception in 1999. Post-war treaties signed in 1947 and 1948 with the Soviet Union included obligations and restraints on Finland vis-a-vis the USSR as well as territorial concessions by Finland; both have been abrogated by Finland since the 1991 dissolution of the Soviet Union. In all this time, Finland has managed to build an extensive welfare state as well as a parliamentary republic premised on equality and egalitarian political culture. Finnish women became the first in the world to win seats in Parliament in 1907; but today Finland can boast being the first European country and the third in the world (after New Zealand and Australia) to allow women to vote in national elections.

The sense of political and social equality that cuts across gender and socioeconomic categories would, invariably, lay the foundation for the brand of consensual political culture characteristic of the Finnish parliamentary system. It is a political culture premised on the principle of communitarian justice—in which justice as a political conception is practical, not metaphysical and epistemological, and is one that can serve as a basis for informed and willing political agreement between citizens viewed as free and equal persons.[14] Today, women have occupied various leadership positions both in the Parliament and in other major institutions of the national government. This norm extends even more to the private sector and in most other social enterprises.

The Finns engage in political and parliamentary discourse, not necessarily with the expectation of embarking on a zero-sum trail of arguments and counter-arguments, but rather with an implicit understanding that a "political conception need not be an original creation but may only articulate familiar intuitive ideas and principles that fit together in a somewhat different way."[15] It is, therefore, left for the people or their representatives to "organize these familiar ideas and principles by means of a more fundamental intuitive idea within the complex structure of which other familiar intuitive ideas are then systematically connected and related."[16] It is by collectively thinking through issues or problems that a fuller understanding of their broader implications and appeal can be made more manifest and profound. By seeking multiple and varied perspectives, Finnish consensual politics "provides a reasonable way of shaping into one coherent view the deeper basis of agreement embedded in the public political culture of a constitutional regime and acceptable to its most firmly held considered convictions"[17]—regarding equality and justice.

It is an acknowledged fact that Finland has for over 100 years been a society in which equality is valued. Originally, this stemmed from the fact that with a small population of about 2 million (at that time), the citizens wanted to bind together and to appreciate each other's mutual existence. Furthermore, the country has not had a history of slavery or involuntary servitude, hence has generally opted for a more egalitarian, classless and non-elitist society. As further validation of these cultural norms, one can look at the non-formal but deliberate processes of facility use within the Parliament. The main cafeteria of the Parliament is accessible to all and is also the same one used by the Prime Minister, members of Parliament, the maintenance staff, the cleaners, janitors and even the cooks and service personnel.

Behind the rostrum of the main Plenary Hall are five bare-naked gilded bronze sculptures of male and female figures with all the anatomical qualities fully depicted. Despite the obvious sensual nature of the display, to which the Finns, curiously, seem quite oblivious, it offers a powerful mirror into the idealistic world of Finnish society and political culture. These sculptures collectively known as "*Work and the Future*" were the works of Wäinö Aaltonen (1894–1966), a leading artist of the time, which was completed in 1932. The four male sculptures symbolized "*Settler, Intellectual Work, Faith* and *Harvester,*" while the female sculptor symbolized the *Future.* Alternatively, these are also known as *The Pioneer*, the *Toil of Thought*, *Faith,* the *Harvester* (representing men's work in society) and the *Future*. The female figure known as the *Madonna* standing with her back to the hall carries the "*future*" on her lap, signified by a small boy.

But the most important relevance of this is that the "changing of Parliament's most important allegory from the *Patria* to the *Future* shows that the Maid of Finland which had fought for law and country during the period of autonomy, had borne a child and become a mother—hence the symbol of young Finnish independence became a maidenly mother."[18] Although Finnish women had received full political rights more than 25 years earlier, the sculptures represent an enduring affirmation of the basic Finnish values of a patriarchal agrarian society committed to work (as a way of transforming its environment) and to the future (a promise of hope, continuity and faith). In this regard, one could ask the question, why are women so much more appreciated and in various leadership positions in Finland's government? The answer is, very simply, because for the Finns women are the mothers of the 'future' of the country.

Identity

In his *Identity and Violence: The Illusion of Destiny*, the Nobel Laureate Amartya Sen wisely points out that "if a sense of identity leads to group success, and through that to individual betterment, then those identity-sensitive behavioral modes may end up being multiplied and promoted."[19] Whether as a matter of reflective choice or as a consequence of evolutionary selection, identity matters in human action. In practice, the Finnish identity is an integral part of Finnish culture. It is a culture that has evolved over the years, first out of necessity but also as a means of adaptation to the enduring challenge of human and social development. As an original agrarian economy dependent on the harvesting of forestry products for domestic and industrial applications, the Finns came to a quick realization that living out of the woods meant that one has to be creative in dealing with nature; hence, the common feeling that Finland was built through self-sacrifice as opposed to some kind of grand Marshall Plan.

Finnish identity, therefore, developed around the idea that that they have paid their own debts to the world. The core of this identity is the premise that 'if there is a problem, then we must do something about it.' It is not a matter of forming committees, but rather, is one of taking practical steps to deal with the problem. It is a matter of 'straight-forwardness' about how to organize minds so as to get very quickly to the crux of the problem. There is, therefore, the tendency to see things in all their potential ramifications, with the caveat that 'if you have to change something, then do it completely.' In the collective sense of the term and seen in various different ways, identifying with others in the community can be extremely important for living in a society.[20]

Industrialization started in Finland from timber and pulp processing. The export of tar and timber products to Great Britain and Germany brought in a lot of money that was used to build and diversify economic productivity. (In fact, Finnish tar was used on the nuts and bolts in London's Trafalgar square). However, after the payment of "war reparations" following the Winter War with the Soviet Union (1939–1940) including the ceding of the Karelian Isthmus to the Soviet Union, the Continuation War (1941–1944) forced Finland to cooperate with Nazi Germany against the Soviet Union. Prevailing accounts see this as an opportunistic marriage of convenience born out of the necessity for the survival of Finland, at a time when military help was coming from no other place.[21] While Finland assisted Germany in its war effort, it also allowed German troops to be stationed in Rovaniemi (a city in the northern part of Finland right on the Arctic Circle). Eventually, and under pressure from the Soviet Union, Finland had to take up arms (the Lapland War 1944–1945) to dislodge and expel the German troops stationed on its northern soil.

Contrary to the conclusion of the Paris Treaty of February 10, 1947, which implicitly indicted Finland of having been an ally of Hitlerite Germany and at war with the "Allied Powers/United Nations," the general Finnish account has been that the country was never politically connected to the ideals of Nazi Germany nor did it embrace its Aryan race policies. It argued that for Finland, its military and foreign policy relationships with Germany were simply a matter of political and military expediency, which was necessary for its survival as a nation in the face of recurring Soviet aggression. As a way of dealing with the financial cost of the post-war reparations imposed on Finland, the shipbuilding and heavy metal industries began to expand. But no matter which account of the circumstances prevails, there still exists among the Finns a seething undercurrent of resigned indignation of having to be unfairly linked (even if remotely) to the dark forces of Nazi military brinkmanship by "misfortunes" beyond the country's control.

The end of World War II was a watershed moment for the evolution of the Finnish state. By 1952, Finland had paid up all the financial bills associated with the war reparation to the Soviet Union, and in the same year it held the Olympic Games in Helsinki. The nascent drive for new technologies climaxed in the rise of Nokia in the 1990s. Nokia, the technology giant was not just good for economic development, it also opened Finland up to the European Union while re-affirming its European identity. Besides projecting Finnish technology and innovation to the rest of the world, Nokia propelled the export economy to an enviable global position.

When the discussion for Finland's accession to the EU began, the country argued that it wanted to be a "buyer of technology" and would like to wait until it had instituted its Faculty of Biological and Environmental Sciences (Life Sciences). It was an early recognition that the paper and pulp industry was losing its economic luster and that biotechnology was in the upswing. Beginning in 1980, the country's investment in research and development as a percentage of GDP rose steadily to 3.7% such that by 2010, it had surpassed almost all the industrialized countries in the world ranking second only to Israel.[22] Because it already had expertise in building ice-breakers and in understanding the behavior of metals in cold weather and experience in working in the Arctic environment, Finland led the effort in the initiation of the Northern Dimension. While economic rationale for the Northern Dimension in the European Union was that it would bring Russia closer to Europe, Finland also hoped that it would strengthen its security relations with Russia by making its (Russia's) foreign policy more predictable.

Cultural Foundations for Development and Social Innovation

Based on the aforementioned discussion, three basic questions can be gleaned: How has Finland utilized technology to sustain a stable welfare state? How does technology transcend existing cultural practices? How do these cultural practices influence civic responses to policies and programs from political and administrative institutions? While these questions are relevant to this work, one of the most recognized studies on national cultures is Geert Hofstede's Culture Index[23] based on six dimensions of power distance (PDI), individualism/collectivism (IDV), masculinity/femininity (MAS), uncertainty avoidance (UAI), long-term/short-term orientation (LTO) and Indulgence (IND).

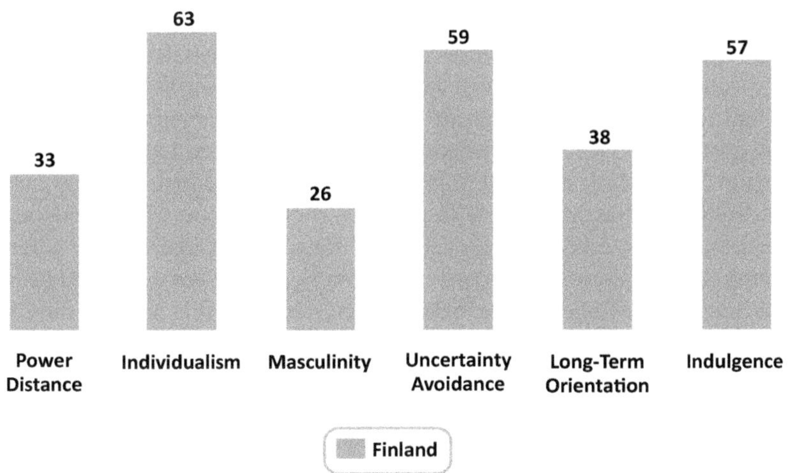

Figure 1.1 The Geert Hofstede National Culture Index.

As a way of obtaining a general overview of the deep drivers of Finnish culture, I draw on further elaborations offered by the Hofstede cultural index, courtesy of the Hofstede Center (www.geert-hofstede.com). Accordingly, power distance (PDI) reflects the extent to which the less powerful members of institutions and organizations within a country expect and accept that power is distributed equally. While Finland scores low on this dimension (33), it reinforces the principles of equality of rights and decentralization of power among the citizens, has less reliance on hierarchy, as well as employee empowerment and a system where control is disliked and attitude toward managers are informal and participative.

The cultural dimension of individualism (IDV) reflects the degree of interdependence a society maintains among its members. In individualist societies people are supposed to look after themselves and their direct family only, but in collectivist societies people belong to "in groups" that take care of them in exchange for loyalty. Finland with a score of 63 is an individualistic society where there is a high preference for a loosely knit social framework in which individuals are expected to take care of themselves and their immediate families, a society that emphasizes meritocracy in hiring and promotional decisions. On the masculinity/femininity dimension, a high score (masculinity) suggests that the society will be driven by competition, achievement and success, with success being defined by the winner—a value system that starts in school and continues throughout one's career. On the other hand, a low score (femininity) means that the dominant values in society are caring for others and quality of life. A feminine society is one where quality of life is the sign of success and standing out from the crowd is not admirable. Finland scores low (26) on this dimension and is thus considered a feminine society. In feminine societies, managers strive for consensus, people value equality, conflicts are resolved through negotiation and compromise, focus is on well-being and status is not shown.

The dimension of uncertainty avoidance (UAI) has to do with the way a society deals with the fact that the future can never be known and the decision whether to try to control the future or just let it unfold of its own accord. The UAI is the extent to which the members of a culture feel threatened by ambiguous or unknown situations and have created beliefs and institutions to try to avoid them. While this ambiguity brings with it anxiety, and perhaps dread, different cultures have developed different ways to deal with it. Finland scores 59 on this dimension and thus has a medium high reference for avoiding uncertainty. Countries exhibiting high uncertainty avoidance maintain rigid codes of belief and behavior and are intolerant of unorthodox behavior and ideas. In this culture, there is an emotional need for rules (even if the rules never seem to work), time is money, people have an inner urge to be busy and work hard, precision and punctuality are the norm, innovation may be resisted and security is an important element in individual motivation. The remaining dimension of long-term orientation (LTO) can be interpreted as dealing with society's search for virtue. It is the extent to which a society shows a pragmatic future-oriented perspective rather than a conventional historical short-term point of view. Finland scores 45 on this dimension thus making it a short-term orientation culture. Societies with a short-term orientation generally exhibit great respect for traditions, a relatively small propensity to save, strong social pressure to keep up with others, an impatience for achieving quick results and a strong concern for establishing the truth. Indulgence (IND) reflects a society that allows relatively free gratification of basic and natural human drives related to enjoying life and having fun. On the other hand, restraint stands for a society that suppresses gratification of needs and regulates it by means of strict social norms.

A basic assumption concerning cultural dispositions is that they change over time, at most retaining their inner and enduring characteristics. While a particular culture will be based in the unique history of a particular group of people coping with a unique set of physical, social, political and economic circumstances, once individuals come to share some set of ideas and practices, these ideas and practices begin to have a life of their own enduring long after the circumstances that gave rise to them are no longer present.[24] The Finnish culture should be seen as a product of historical experience and coincidental forces that have resulted in a tradition and habit that remain resilient even in the face of competing modernity. As a people and as a culture, the Finns rarely had it too easy, but the lessons learned (good or bad) were employed as not a defeat but a positive challenge to do better.

By virtue of its relative geographical isolation and harsh weather conditions, Finland nonetheless was endowed with enormous forestry and wood products. It excelled in agriculture, food productivity, mining and manufacturing, which became the initial drivers of its export economy. By shaping industrial productivity in line with the comparative advantage offered by nature and the environment, the country was able to mobilize the human and capital skills necessary for development. Through technology and innovation introduced and supported as a matter of national policy, it was able to establish transformative industries that further strengthened the export economy in terms of opening new trans-European markets and value-added production. Not only did the productive culture rest with the economy, it also transitioned into the larger society and improvements in the quality of life and social welfare with particular emphasis on 'softer innovations'—those that help society function better, such as universal education, social and health systems.

A Pioneering Attitude toward Innovation

Finland built much of "her successes on cumulative knowledge and skills in creating the new, relying on people's ability to learn, and respecting the institutional memory—especially where tasks and ways of doing things in society are concerned."[25] These represent valuable knowledge based on broad competencies that have slowly matured over the years. Kyösti Karjula[26] succinctly illustrates the ideational foundations that undergird the Finnish disposition toward enduring innovations. According to him in the following explanations, the first was the Finnish faith in knowledge and consequent investment in education. As long ago as the 1840s, the Senate attempted through the newly established "Manufacturing Board" to encourage Finnish entrepreneurs to be interested in new inventions and patents. It also sent talented students abroad to institutions of learning—thus making Finland a major benefactor of what international scientific and professional organizations had to offer. Imitation and following in the tracks others have paved thus became a good operational model.

Fate and Providence have also created waves of learning in the country at specific historical junctures, and the results have taken time to manifest themselves over the years. The first wave was centuries ago during the period of Swedish rule (1155–1809); the most important decision then was that people would not be granted permission to marry unless they could read and write. Although that decision made Finland a country of literate people, it did not solve the persistent problem of poverty and underdevelopment. The next wave of learning began in the nineteenth century but in practice achieved its goal only in the years after the country gained independence from Russia in 1917. The state decided that every boy and girl irrespective of socioeconomic status must attend school—in that way everyone would be given the opportunity to succeed through education. These two initiatives reinforced two normative principles that would eventually pay off for Finland in dramatic ways—the values of an equitable egalitarian society and opportunity for success open toall.

The third major wave in social and economic renewal took place after World War II in the 1960s and 1970s. "By creating the nine-year comprehensive school, which all children could attend, the door for opportunities for a better future was flung wide open for the entire nation."[27] In recognition of the fact that, "laying the groundwork for a robust economy also demanded a reform of the higher education system, two critical decisions that would stake out the course that Finland would take for many years into the future were made: government loans for studying were created, and university education was decentralized throughout the country by founding tertiary institutions in regional centers."[28] These policies were instrumental as the country made the transition to a competence -based society as would later be demonstrated in her performance at the head of the Programme for International Student Assessment (PISA) statistics.[29] By building an effective infrastructure of public-private partnership through the Finnish Funding Agency for Technology and Innovation and SITRA—the Finnish Innovation Fund, the country was able to build a collaborative network that allowed university research and scientific discoveries to be transferred to manufacturing industries where they are turned into functional products for the domestic and export markets.

A befitting summation would concur with the words of Kyösti Karjula that Finland as an economy has in 100 years accomplished one of the most impressive leaps that any country has ever made from being one of Europe's poorest to one of its wealthiest. Wars have not prevented the country from prospering, and every generation has doubled its real income per capita. The accumulated wealth is not only material but also non-material. The country today has a functional democracy, equality and a dependable administration that would provide a good institutional leadership into the future. This is not to say that the country expects the future to be all rosy and predictable. Policy makers understand that the best laid plans may, after all, not come to fruition at the right or anticipated moment and that today's promise and opportunities have a curious way of turning tomorrow's vision into a seething web of pessimism and melancholy. It will take a determined people and a great effort to overcome the challenges of the future.

Notes

1 http://www.newworldencyclopedia.org/entry/Finland; http://en.wikipedia.org/wiki/Finland. Accessed on August 15, 2015.
2 Suvikumpu, L. 2010. *Find out about Finland*. Keuruu, Finland: Otava Publishing Co. Limited, p. 26.
3 Kalu, K. N. 2011. Institution-building, not nation-building: A structural-functional model, *International Review of Administrative Sciences,* 77(1): 119–137, 126.
4 Trice, H. M. and J. M. Beyer. 1993. *The culture of work organizations*. Upper Saddle River, NJ: Prentice Hall.
5 Leidner, D. E. and T. Kayworth. 2006. A review of culture in information systems research: toward a theory of information technology culture conflict. *MIS Quarterly,* 30(2): pp. 357–399.
6 Keesing. R. M. 1974. Theories of culture. *Annual Review of Anthropology,* (3): 73–97.
7 Statistics of Finland. 2013. Bureau of Economic Analysis, September 30.
8 Nykanen, P. and H. Paulapuro. 2005. *Around the roll: A century of Finnish paper machine and papermaking technology*. Helsinki, Finland: The Society for History of Technology, p. 13.
9 Ibid. p. 15.
10 Ibid. p. 27.
11 Ibid. p. 31.
12 Trice, H. M. and J. M. Beyer. *The culture of work organizations*.
13 The US State Department. 2007. See http://www.infoplease.com/country/profiles/finland.html. Accessed on August 15, 2015.
14 Rawls, J. 1992. Justice as fairness: Political not metaphysical. In *Communitarianism and individualism*, ed. S. Avineri and A. de-Shalit, 186–204. New York, NY: Oxford University Press, pp. 192–194.
15 Ibid. p. 192.
16 Ibid. p. 193.
17 Ibid.
18 Parliament of Finland. 2000. *The voice of the people: Past, present, and future*. Helsinki, Finland: Parliament of Finland, p. 219.
19 Sen, A. 2006. *Identity and violence: The illusion of destiny*. New York, NY: W. W. Norton & Co, p. 23.
20 Ibid. p. 19.
21 Wuorinen, J. H. 1965. *A history of Finland*. New York, NY: The American-Scandinavian Foundation.
22 Eurostat, OECD 2013.
23 The Hofstede Center, Dimensions of National Culture. www.hofstede.com. Accessed on May 1, 2015.
24 Trice, H. M. and J. M. Beyer. *The culture of work organizations*.

25 Karjula, K. 2006. To the reader. *Finland, decide to succeed! A background polemic on innovation policy*. Helsinki, Finland: Secretarait of the Committee for the Future, The Eduskunta, p. 5.
26 Ibid. pp. 22–23.
27 Ibid.
28 Ibid.
29 Ibid. p. 23

2 The State, Parliamentarianism, and Public Policy

The evolution of Finland's political history was shaped in fundamental ways by its having been ruled by Sweden for almost 700 years (1154–1809). Although under the authority of Sweden, the Finns for the most part always sought to maintain a certain level of sovereign and independent national aspiration. It was also during this period that a representative decision-making system was created, from which parliamentary democracy based on universal suffrage would later develop in Finland.[1] Over the years, a series of opportune events continued to shape the trajectory in which Finnish politics would evolve. In July 1788, King Gustavus III declared war against Russia without the approval of the Swedish Parliament, the *Riksdag*. Military officers numbering up to 113 (which included both Swedish and Finnish officers) under the platform of the *Anjala League* objected to the war saying that Sweden had begun an aggressive war against Russia, hence "the conviction of the inevitability of war which creates in a people born to freedom, fighting men who cannot be defeated had vanished."[2]

Although the war against Russia ended with the signing of the Peace of Varala on August 14, 1790, and a return to the status quo ante, the reaction of the *Anjala League* to the war, though seen by many not as an expression of nationalist purpose,[3] may have added increased resonance to a simmering but hitherto unpopular Finnish "independence" aspiration. Prodded by Napoleon to invade England following the long wars of Revolutionary France, which began in 1792, Alexander I of Russia declared war against England on November 11, 1807, and invaded Finland (still a Swedish estate) on February 21, 1808. The war ended in favor of Russia and led to the "signing of the treaty of Fredrikshamn on September 17, 1809 (almost two months after the end of the Porvoo Diet); Sweden formally ceded Finland to Russia, including the Åland Islands, which had long been an administrative part of the Finnish half of the Swedish Kingdom."[4] However, rather than outright annexation or subjugation, Alexander I sustained a "different arrangement which enabled Finland to retain her pre 1808–1809 constitutional, legal, administrative, and religious institutions with little or no fundamental change; such that as part of the Russian Empire, Finland retained the status of a self-governing, constitutional state, with all the basic instrumentalities essential in such a state."[5]

A watershed moment came when, at the instigation of Alexander I, a meeting of a few representatives of the Four Estates of Finland was called in St. Petersburg, Russia, as a way of discussing the needs and wishes of the Finns under Russian rule. This resulted in the convening of the first Finnish Diet (The Porvoo Diet—a de facto parliamentary legislative body) made up of representative members of the Four Estates (the *nobility*, the *clergy*, *burghers* and 'land owning' *peasants*) who met in Porvoo on March 25, 1809. ('Ordinary' Finnish peasants were not represented in the Diet until 1563). The result led to Russia's acceptance of the tenets of the *Act of Assurance* to maintain the constitution and laws of Finland, as well as a tacit approval of a self-governing Finnish state. On April 4, 1809, both the allegiance of the four estates and the *Act of Assurance* were promulgated into a solemn decree by Alexander I (now considered a constitutional monarch with the title of Grand Duke of Finland as well as Tsar of Russia). Even under Russian imperial rule, Finland's democratic potentials were quite limited. Much of the basic structure of its laws were still imbued with characteristics of Sweden's Law of 1734 and royal absolutism on the one hand and on the other the political whims of Russia's Tsar Alexander I, who could be said to have guaranteed Finland's political existence and the maintenance of law and order but not the government deemed suitable to the needs and legislative aspirations of the Finns.

2.1 Primordial Nationalism

While local sentiments endured for most of the period under Russian rule, "the severance of the centuries-long connection with Sweden also caused a quickening of Finnish patriotism and contributed to a national awakening destined to carry the nation far from the moorings of 1808–1809."[6] "The strengthening of love of country became part of the development of a Finnish national consciousness; local and provincial loyalties had to be flattened out and make room for the larger concept of people, nation and state."[7] The surge of national consciousness was reinforced by a rising sentiment among the educated classes who, being more cognizant of the signs of the times, understood its general implications, particularly the union with Russia.

The earliest overt expression of nationalist views began with the works of Adolf Iwar Arwidsson,[8] a poet, historian and newspaper publisher who saw a common Finnish language as the foundation for a Finnish national identity. The result of Arwidsson's advocacy led to progress in the field of historical studies and the teaching of national history, the need to introduce on a larger scale the study and use of Finnish as the language of instruction in schools and the compilation and publication of a growing body of folklore. The accomplishments of patriot-statesmen such as Johan Wilhelm Snellman further energized the Finnish nationalist movement. "In 1828, the first lectureship in Finnish was established at the university, and in 1843 Finnish was made a part of the curriculum of the secondary schools. The first professorship in Finnish was established at the University of Helsinki in 1851, with the incumbent being M. A. Castren, and the chair was assigned to Elias Lonnrot, the physician-folklorist in 1853."[9]

Among the men who used their writings on Finnish folklore, stories and history to drive the national consciousness[10] were Zachris Topelius (1818–1898) whose leading works on Finnish history and patriotic literature (*The Book about*

Our Country, and *The Tales of an Army Surgeon and Stories for Winter Evenings*) charted the course on what it meant to be a Finn; Elias Lonnrot (1802–1884), who wrote the national epic and folklore *The Kalevala* (1835); and Johan Ludvig Runeberg (1804–1877), who published *The Elkhunters* (1832), a long hexametric poem describing the life of the humble farmer in central Finland, and the *Tales* (1860). In this regard, the drive for Finnish nationalism and statehood was spearheaded not by a mass movement of frustrated peasantry, but rather by a committed visionary elite. Neither was there a need to arouse a level of mass consciousness, but it was imperative to educate the body of Finnish citizens on the path for which Providence has prepared for them and for which they must pursue. The efficacy of the folklores, national myths, and historical anecdotes therefore lie in their serving this purpose, while "at the same time, removing a rich body of meaning from the field of explicit statements, to the point where the future course of history would have to be revealed in symbolic form"[11]—the state of Finland.

This factor was even more accentuated as a result of the second Diet of 1863, convened at the Imperial Palace in Helsinki by Tsar Alexander II, as well as the Language Decree of 1863, which sought to make Finnish one of the official languages of Finland in addition to Swedish. The Diet of 1863 encouraged increasing and more frequent legislative work and spurred the formation of political groups and publications that cut across Estate lines. "The Municipal Acts of 1865 and 1873 further increased the opportunity for citizen participation and influence in decision making, such that by the 1890s, interaction between civil society and the Diet's legislative work had taken on such established forms that this led to more open debate concerning social alternatives."[12] By 1906, Finland was able to create a unicameral Parliament based on universal suffrage, and in 1907 it held its first democratic parliamentary elections (even while still under Russian rule) in which women were first allowed to vote and also to stand for public office. In addition to the rise of the Social Democratic Party, other political parties with competing persuasions were also formed, but "the rise of multi-party system in addition to the expanded right of expression and frequent parliamentary elections drew attention to increasingly sharp divisions along ideological lines."[13] This very fact, in addition to the characteristic norm of seeking agreement on very tense issues, has meant that Finnish legislative politics is premised on respect for the rule of parliamentary democracy, making political compromises and governance through coalition politics—involving political parties with quite different ideological orientations.

In the end, it can be generally surmised that, in all, three cross-currents of events added momentum to the appeal of independence. First was the increasing attempt at *Russification* of Finnish political and administrative life "and an Imperial manifesto issued on February 15, 1899 that placed all Finnish legislation under the surveillance of the Russian Government, and in which the constitution and laws of Finland would be interpreted in light of Russian and not Finnish interests."[14] The second was the onset of World War I, which invariably provided Russia the incentive to apply increased dictatorial and arbitrary rule over Finland amid great resentment; the third was the event of March 1917 that brought about the Bolshevik revolution in Russia and the consequent collapse of the autocratic regime of Tsar Nicholas II.

2.2 Parliamentarianism: Toward Constitutional Democracy

Amid increasing uncertainty about the potential for anarchy in Russia following the fall of the Tsarist regime, the Finnish parliament on November 26, 1917, set up a non-Socialist coalition government headed by Pehr Evind Svinhufvud. Parliament also approved the declaration of independence on December 6, 1917, and Svinhufvud (November 27, 1917 - May 27, 1918) became the first head of Cabinet and Prime Minister of an independent Finland. The ensuing civil war (1918) between the Bolsheviks (Reds) and the Whites ended in the defeat of the Reds but left a bewildered national spirit seeking answers and asking questions at the same time. On July 17, 1919, Parliament approved a constitution establishing a republican form of government with a unicameral multi-party parliamentary legislature. The President was chosen by a legislative vote of 143 out of 200, which saw Kaarlo J. Stahlberg emerge as the first president of an independent Finland (1919–1925, See Appendix A). Legislative power is vested in the Parliament, while executive power rests with the Cabinet (Council of States) headed by the Prime Minister who is the Head of Government. "The constitution also provides for the president to be chosen for a six-year term by an electoral college of 300 which in turn was elected by universal suffrage. It also provided for an independent judiciary, and a bill of rights guaranteeing various civil rights and freedoms."[15]

Based on the 1919 Constitution, which lasted until 2000, "foreign policy was made the preserve of the President, who also had the right to appoint the Government and to dissolve Parliament and call for new elections. As further checks and balances, members of Parliament can present arguments and move for a vote of no confidence in the Government; and by a majority vote of its members, Parliament could force the Government to resign."[16] Some procedural changes have been made following the passage of the new Constitution of 2000, which, invariably, increases the power of the Parliament in relation to the government and the power of the government in relation to the President. Today, although members of Parliament conduct government negotiations independently, the President nominates the Prime Minister, but the Parliament must ratify the nomination with an absolute majority vote. Additionally, the Prime Minister nominates the remaining members of the Parliament who then with the consent of the Parliament (Council of State), are formally appointed by the President. "In the end, the Prime Minister is the only person with the power to propose that Parliament be dissolved."[17] "Members of Parliament are elected for a four-year term by proportional representation from 15 multi-member electoral districts under universal suffrage at age 18; while the president (Head of State) is elected for a six-year term."[18] Today, Finland has a robust parliamentary democracy but one that is always seeking to bring new knowledge and innovation into the craft of legislative politics and democratic accountability.

The rest of this paragraph offers,[19] in a nutshell, the contemporary basis in design, structure and processes of presidential and parliamentary politics. Constitutionally, the 200-member, unicameral Parliament (*Eduskunta*) is the supreme authority in Finland. It may alter the constitution, bring about the resignation of the Council of State and override presidential vetoes; its acts are not subject to judicial review. Legislation may be initiated by the president, the Council of State, or one of the Eduskunta members. The Eduskunta is elected on the basis of proportional representation. All persons 18 or older, except military personnel on active duty and a few high judicial officials, are eligible for election. The regular parliamentary term is four years; however, the President may dissolve the Eduskunta and order new elections at the request of the Prime Minister and after consulting the speaker of Parliament.

Finland has a mixed presidential/parliamentary system with executive powers divided between the President, who has primary responsibility for national security and foreign affairs, and the Prime Minister, who has primary responsibility for all other areas, including EU issues; much of this is reinforced in the 2000 Constitution. Under the constitution that took effect in March 2000, the established practice for managing foreign policy is that the President keeps in close touch with the Prime Minister, the Minister for Foreign Affairs, and other ministers responsible for foreign relations. Constitutional changes strengthened the Prime Minister—who must enjoy the confidence of the Parliament (Eduskunta) at the expense of the president. The President:

- Is elected directly by the people for a six-year term; since 1991 no President may be elected for two consecutive terms of office;
- Handles foreign policy, except for certain international agreements and decisions of peace or war, which must be submitted to Parliament, and EU relations, which are handled by the Prime Minister;
- Is Head of State and Commander-in-Chief of the Armed Forces and has wide decree and appointive powers;
- May initiate legislation, block legislation by pocket veto and call extraordinary parliamentary sessions; and
- Appoints the Prime Minister and the rest of the Cabinet (Council of State). The Council of State is made up of the Prime Minister and ministers for the various departments of the central government as well as an ex-officio member, the Chancellor of Justice. Ministers are not obliged to be members of the Eduskunta (*Parliament*) and need not be formally identified with any political party.

The President may, upon proposal of the Prime Minister and after having heard the parliamentary groups, order parliament be dissolved and a new election held. The multiple consultations, though often tedious, help to reinforce the consensual character of Finnish legislative politics. As illustrated in the *Encyclopedia of the Nations*,[20] since 1945 no single party has ever held an absolute parliamentary majority, so all Cabinet or governmental decisions involve coalitions.

The Cabinet is composed of the heads of government ministries and has as its primary responsibility the preparation of governmental budgets and legislation and the administration of public policies. The Prime Minister and Cabinet serve only so long as they enjoy the support of a working majority in Parliament, and there have been frequent changes of government. On gender equality, women are fairly well represented in both the executive and legislative branches of government in Finland. Women hold 38% of the seats in the 200-member Parliament (Eduskunta), and there were eight women among the 18 cabinet members in 2003. Women have held top leadership positions, including Defense Minister (Elisabeth Rehn), Foreign Minister (Tarja Halonen) and Speaker of Parliament. In February 2000, Finns elected Tarja Halonen their first female president. Anneli Jäätteenmäki was named prime minister following elections in March 2003.

2.2.1 The Allure of Consensual Politics

Besides helping to build civic trust in the political system, consensual politics is a way of actualizing the core principles of an egalitarian social welfare system that has been a major driver in collective public service. Although there were considerable differences between parties concerning how quickly the nation should proceed in creating a broad social insurance system, extensive social and health care and a comprehensive school system, political differences were smoothed out by broad-based Government coalitions that cut across left, right and center-leaning parties[21]—a feat rarely seen in the western capitalist world. There is another factor that seems to drive Finland's proclivity toward consensual politics. The civil war (1918) was a watershed moment that was quite impactful in reshaping Finland's legislative political culture, the consequent evolution of the liberal state and the brand of consensual democracy that has characterized policy making in the Parliament. The lasting experience of the event in which Finns set upon each other shook the national psyche, such that even in the midst of ideological disagreements between the republican "Whites" and the Bolshevik "Reds," the experience would serve as a unifying force as Finns collectively fought against the Soviet Union in the consequent Winter Wars (1939–1940). On various accounts, both natural and man-made, Finland is a state born out of necessity and a practical realization that human intelligence should be taxed to find peaceful means of mutual co-existence between people. Life as a higher calling offers a rational basis for bridging differences among contending parties. Therefore, the Finns see consensual politics as a means of practical decision making but also as a means of healing and reinforcing its anti-elitist egalitarian social sentiments.

The Finns put a lot of thinking into almost everything they do. Even the aesthetics and architectural themes within and outside the Parliament building were not spared. The construction on the cube-shaped granite building that houses the Parliament began in the 1920s and continued until it was inaugurated in 1931. The architect was a young Finn named Sigfrid Siren (1889–1961). The building is deliberately "eclectic in its architecture and decoration, with modernist elements of functionalism applied in a classical framework; features of art deco, which was influenced by cubism and oriental exoticism, stand out clearly in decorations and textiles."[22] "In its stately solitude on Arcadia Hill this edifice is at the same time calmly monumental and nearly defiant in its grandeur."[23] Besides its significance in strengthening national pride, the building was a showcase for Finnish interior design and crafts, which helped form the young nation's public image, freedom and independence.[24] Bedecked in finely grained, redish *Kalvola* granite, the muscular-looking exterior of the building is a light, neutral beige; the same is true for the inside interspersed with hues of mild soft colors. There is no red (representing images of *blood* and *violence*) or green or white colors (which generally could stand as symbols of one *dogma* or the other).

Within the building, there are no art works or effigies that personify war, victory or conquest. Even the ubiquitous statue of *Lady Justice* that has been a common sight in many legislatures or court systems carries a scale in one hand, but no *sword* in the other. Every design, craft, decoration and motif has a historical meaning. The Finns mourn the "mistake" of the civil war and their self-indignation and atonement is symbolized by the sculpture of a bowed *Lion of Finland without Its Crown,* which is located on the wall of the Parliament's cafeteria. This was the work of the famed sculptor Gunnar Finne (1886–1952). The plaintive allegory of a female kneeling beside and comforting a bemused lion is quite striking and profound while it also resonates a deeply felt regret and a plea for national healing from the travails of the civil war.

2.3 Egalitarian Democracy: Symbolic yet Significant

The location of Finland's parliamentary building on Arcadia Hill in Helsinki is not only unique but also quite symbolic. Unlike parliamentary or legislative buildings of most countries, where you have fences, wide expanses of land and some distance separating the building from the closest civilian or public infrastructure, Finland's parliamentary building is situated more or less in the middle of Helsinki—a stone's throw from the central railway station in Kluuvi (central Helsinki). It is bordered on all sides by roads and pedestrian walkways, private businesses, the Museum of Natural History, coffee shops and cafés, boutique shops and corporate offices. Pedestrians and drivers go by the building sometimes unmindful of but unmistakably recognizing its presence.

Even with its stern-looking façade, the building projects a psychological aura of "fitting in" with the surrounding aesthetics: there are no overt displays of gun-toting security personnel; security cameras are there but are less obvious to the public. The Finns see legislative work as part of the people's work; hence, the building should also be part of the environment in which the people live and work. It casts a friendly and welcoming disposition as opposed to the fortress mentality one would experience in most legislative buildings around the globe. Furthermore, it allows legislators and public servants to stay in close touch with as well as feel the pulse of the citizens on a routine basis—a continuous validation at least in theory of the central democratic axiom that power comes from the people, and legislators are there to do the people's bidding.

Ironically, one obvious question arises from this observation: To what extent does the average Finn feel connected with or at least remain independently judgmental about the current temperament in Finland's politics and decision making? While most remain politically aware, there exists a sense of benign detachment from the daily routines and pragmatics of the political world in which they live. For most people, a quick reference check of the headlines would be enough. Public anxiety is drawn out more on sensational and crises-driven issues rather than the mundane and routine. It is not that the average Finn does not trust his or her legislators or the government itself; it could be that he or she very much trusts the government and its functionaries to always do the right thing and this trust causes citizens' attention and critical analysis to shift to other matters. "Although voter turnouts in Finland are high, but only forty percent or so say they have at least some degree of confidence in the parliament"[25] and "many feel that the privileged are still giving the poorest only enough for indispensable subsistence out of public funds."[26] This could be like the situation in the United States where the polls consistently indicate that a majority of citizens do not trust Congress as an institution but trust their own individual congressman or legislators.

As a result of the broad application of information technology in Finnish public and administrative culture, the people are able to receive a lot of information, at times more than is desirable. The tendency for information overload creates a situation where selective attention and catchy headlines become the rule of the game. They become coping mechanisms that allow citizens to manage the deluge and overcrowding of information they receive on a daily basis. In this regard, most mundane legislative issues are given less attention or critical review than they deserve; on other hand, legislators are given a pass because they have become the 'pseudo' experts in specific program areas or policy responsibilities and are thus allowed to decide what is best for the people. After all, is that not the foundation of representative democracy?

With an aging population, many of whom are less attuned to the attraction and demands of utilizing information technology in their daily lives, one therefore wonders how many citizens are actually *tuned in* to what the Finnish Parliament is doing. This problem is made more manifest by the increasing political indifference and secularity of a rising youth population. How, then, do legislators reach out to their constituencies in more robust and meaningful ways? The attempt to bridge the procedural schism between indirect (representative) democracy and direct democracy fails simply because the pattern of communication and information exchange is essentially unidirectional, weighing more heavily in favor of government and its functionaries than the average citizen. By regulating the parameters and context of the message as well as its timing and dissemination, the Parliament, either by design or default, is able to leverage its constitutional status and its control of information to shape the way the message is received and often interpreted by the public. In relative terms, they understand that the average citizen is less likely to devote the time needed for a deeper analysis of crucial public policy issues. While this may not be unique to Finland, it is nonetheless a recurrent theme that resonates in most critiques of representative democracy.

2.3.1 Organizational Learning and Knowledge Management

In representative democracies, particularly the United States, the committee system is the workhorse of Congress, including the various substantive subcommittees. While many committee members do not at the beginning possess the technical expertise or the skill sets needed to engage the analytical requirements of their policy or program jurisdictions, they learn through the process itself, and in the end they become experts. By utilizing information derived from research, crowdsourcing and related testimonies, the committee is able to make recommendations to the whole legislative body, which further engages in a series of debates on the issue. In the end, disagreements, conflict, compromise and accommodation generate a structural template in which learning and innovation occur. The legislature becomes a learning organization, both in the way changes in the legislative process are made and also in the resolution of seemingly intractable policy controversies.

The committee system in the Finnish Parliament is part of a larger learning organization. They deal with crucial political, social and economic problems in ways that seek solutions to contemporary problems of society, sometimes devoid of the kind of legislative hubris and cynicism typical of most parliamentary systems. Two prominent committees are of note: the Grand Committee and the Committee for the Future. The 60-member Grand Committee is the controlling and coordinating organ within the unicameral legislature. It is also charged with playing a central role in Finland's European Union policy (EU)—a role that enables it to play a moderating influence on public sentiment regarding the preservation of Finnish identity and independence as well as policy integration and collective identity in the context of a supra-national European identity. Suffice it then to say that in the context of the European Union, all politics are still national even in the face of a collective EU strategy. Many of the countries remain lukewarm regarding a wholesale embrace of full European integration. Beyond that of sovereign identity, some of the issues seem to impinge on a reluctance to accept a common tax structure among the member states for funding the projects and measures of the EU, as well as differences in foreign policy posture. For instance, Finland is not yet a member of the North Atlantic Treaty Organization (NATO), partly as a result of its neutrality policy and because it does not want to 'offend' or undermine the tense peace with its neighbor to the East (Russia).

The Finns want a peaceful relationship with Russia, but they are not oblivious to the fact that Russia remains a formidable force that must be appreciated, accommodated and if necessary appeased for the sake of peace. While Finland plans for its own defense, it is cautious of the presence of Russia in light of its enormous military strength and rising wealth as a result of oil. Hence, Finland's foreign policy approach to Russia has generally been to pursue its own sovereign and independent national interest, but also be clever in doing so. There is an overwhelming sense of practical realism in Finnish thinking regarding Russia, and this mindset has remained an undercurrent of its foreign policy and international relations, especially in circumstances where its own national interest may seem to have the potential of colliding with that of Russia.

2.3.2 The Committee for the Future: Mitigating Uncertainties and Surprises

A key element in Parliament's legislative work is carried out by an aptly named *Committee for the Future*. While some of the Committee's work is dealing with current domestic policy issues, its activities are central to shaping a more prosaic understanding of political and social changes in Finland and the country's role in the rest of the world. Because it has an "open-ended" mandate, there are no limits to the issues or matters the Committee can discuss or deal with; no one tells the Committee which issues are important. It can institute research on various issues and seek expert advice from within and outside Parliament, from consultants or various research institutes. Most of its research is in cutting-edge work centered on "radical technologies" or top technologies that will *change* the world today and in the future. The Committee sits twice a week; there is no voting among the members, but the Committee's role is to sit down and *think*—about pressing contemporary issues and what the future holds for society and mankind in all of its earthly ramifications.

Although funded by Parliament, it reports to the plenary session; the Prime Minister can also report on the work of the Committee to the Parliament. While the membership changes every four years, the Committee has issued approximately six reports in a little over 20 years. The *Reports* deal with issues including globalization, the future of Europe, aging, population policy, technology and society's evolution, technology's impact on work and work culture, the changing dynamics and problems of the Finnish welfare state, cataclysms and/or *black swans* that will change the world as we know it or any other issue deeper imagination can bring to bear. The Report goes to the Plenary Session of Parliament and is eventually issued as a *White Paper*.

As a central tool of policy making, the Committee's work is both contemporary and futuristic, drawing upon potential but seemingly impossible scenarios that hinge on the unthinkable. But, what if the unthinkable eventually becomes real? How will government and societies deal with it? If we do not have ready-made answers to these possibilities, then it behooves us to, at least, to imagine their occurrence and to have a reservoir of options and alternatives where possible to address or confront worse-case scenarios when and if they materialize. One problem that has been discovered about the Finnish welfare state is that most people involved in it do not want to go back to work because most of their needs are served by the state, and everything else seems to be in order. While the reverse may, in fact, be the case, what should the government do about the rising level of public apathy and indifference to the various challenges faced by society? If left unchecked, how would Finland sustain its current development path and future productivity? Even when there are no ready-made answers, these are some of the problems that the Committee for the Future tasks itself.

2.4 Knowledge Management and Parliamentary Work Processes

Finland is a country that has transformed itself into what is regarded as an information or knowledge-based society. It is believed that such transformation should be reflected in not only the way public services are delivered but also the way decision making and policies that inform those services are handled. Because knowledge management can be seen as a continuous process of learning, adaptation, innovation and change, it can greatly influence how society develops and parliamentary work processes and decision making are designed and framed. At the core of knowledge management is the ability to integrate various facets and manifestations of information, communications and technology (ICT) into a shared frame of reference and functional vision, to define and elaborate on joint objectives, to change the prevailing work culture and to arouse people's interest in vital matters of national relevance.[27]

As a learning tool, knowledge management seeks to enhance Parliament's routine work as well as its capacity to react more swiftly to societal change as a decision maker. In this regard, it helps to facilitate democracy as well as parliamentarianism. It is equally noteworthy that the "central factors causing pressure for change are the management of information overload, and the need for organizations to learn new things in a rapidly changing action environment."[28] In that much of Finnish society spends its time on continuous information processing, an increasing sense of information overload can stunt independent creative thinking. It can disorient work processes and cause much time to be spent collating and arranging several pieces of information from multiple sources into specific clusters of knowledge for use toward desired ends.

While the requisite technology for extensive interaction and communication already exists in Finland, "the next question is how to make the technology serve people, and how to make people utilize the opportunities available."[29] The Parliament's Committee for the Future therefore has sought to find ways to make information and technology more outcome-friendly so that the art of "knowledge management becomes a process where knowledge, skills, expertise and communication are catered for, administered and steered with skill and wisdom in a goal-oriented fashion."[30] A good example of this is the increasing emphasis on *crowdsourcing* as a way of increasing citizen participation, diversity of information and perspectives and citizen empowerment in the policy-making process. The Parliament-wide knowledge management project, launched in the autumn of 2000, seeks to develop four critical value-based *knowledge management principles*[31] for its legislative activities:

- *Trust, positive speculation, and creative processes* Continuous haste is seen as a problem in parliamentary work and cannot allow people to develop appropriate problem-solving skills. While haste cannot be eliminated through technology, it is most important to pay attention to what people want. Creative thinking and the development of jointly produced ideas are central in eliminating haste in parliamentary deliberations.
- *Openness, influence, and developing competencies* Openness is emphasized in the interaction between members of Parliament, between members of Parliament (MPs) and citizens as well as in preliminary legislative work among government officials. Further competencies entail raising the expertise of members of Parliament while also emphasizing their personal mission. Each MP is required to produce his or her personal development

plan in the form of a work chart that requires continuous updates. Concrete partial learning targets would be scheduled in a calendar with enough time reserved for learning. The fact that each MP has a personal mission will enable the construction of a personally significant expertise profile; this would highlight critical elements of expertise rather than forcing everyone to master everything. It encourages cooperation between MPs and the Ministries through improving joint activities during the bill-drafting process. Bills submitted to Parliament would be more illustrative with appropriate audio-visual advancements; developing quality assurance system for legislative work would also allow for consideration of regional differences in the laws being passed and the need to ensure adequate resources for expert group consultations.

- *Sharing, message, and virtual networks* Because parliamentary work is versatile and presupposes prompt reactivity, MPs would be able to operate efficiently in various networks (virtual network activities, e-learning environments or other state-of-the-art technologies). This offers a communication toolbox that could be developed and updated on a continual basis. Acquiring the requisite competencies would require a knowledge toolbox containing the following:
  - Easy-to-use links to existing files that are organized per subject matter
  - Automatic access to updated key numbers, messages and novelties
  - Links to central think tanks and public authorities providing analytic and pre-processed knowledge that the MP in question has selected for monitoring
  - Information service organized per subject matter for use by parliamentary committees
  - Easy-to-use links to existing databases to enable active and pertinent interaction during legislative processes
  - Internet pages and ready-made activity packages to enable debate initiated by individuals, either alone or together with specific groups, concerning critical issues related to committee work or parliamentary processes.
  - Crowdsourcing as a way of engaging citizens in policy making as well as spreading awareness about issues currently in the decision-making process
  - Parsimony in the exploitation of knowledge without a deluge of information by producing service packages of various levels for specific MP needs
- *Doing things together and learning together* As a way bringing together all facets of the knowledge management system, the emphasis is on creating joint activities and learning together among members of Parliament. It also requires the designation of a location and space with sufficient opportunities to generate the type of activity leading to the creation of new knowledge. The idea of learning-oriented discussions among MPs, public employees and experts from various sectors is to create a forum based on MPs' own interest in their line of work and articulation of main trends and weak signals related to national issues. Currently, the Parliament café

> serves the purpose of knowledge distribution relatively well with various parties and sectors meeting to share information and one another's views. Because "one of the most visible challenges of knowledge management is that one of the most valuable sources exist mainly in an undocumented form as tacit knowledge in the minds of individuals, informal meeting places such as the Parliament café helps to create a cultural and technological environment in which all other members of Parliament could exploit this knowledge."[32]

For Finland, what has been gleaned so far is that an effective knowledge management system that draws from both tacit and explicit knowledge could provide an integrated framework needed for success in parliamentary decision making and in the advancements needed for the deepening of democracy. The central element for achieving success in this regard is focused on implementing a lifelong learning strategy, developing national innovation systems, increasing investments in research and development and serving as an information society laboratory within the European Union.[33] Members of Parliament are not only engaged in policy making, they are also engaged in a continuous process of learning, research and discovery. As an institution, the Parliament has become a center of innovation, broad-based consultations and deliberative inquiry.

Notes

1 Parliament of Finland. 2013. *Democracy's long road*, p. 4.
2 Wuorinen, J. H. 1965. *A history of Finland*. New York, NY: Columbia University Press, p. 105.
3 Ibid. p. 108.
4 Ibid. p. 116.
5 Ibid. p. 113.
6 Ibid. p. 136.
7 Ibid. pp. 136–137.
8 Ibid. p. 142.
9 Ibid. p. 143.
10 Ibid. pp. 145–152.
11 Voegelin, E. 1957. *Order and history: Plato and Aristotle*. Baton Rouge, LA: Louisiana State University, p. 151.
12 The Parliament of Finland. *Democracy's long road*, p. 8.
13 Ibid.
14 Wuorinen, *A history of Finland*, 202.
15 The Parliament of Finland, *Democracy's long road*, p. 13.
16 Ibid.
17 Ibid. p. 20
18 *Encyclopedia of the nations*. Finland–Government. http://www.nationsencyclopedia.com/Europe/Finland-Government.html. Accessed May 31, 2015.
19 Ibid.
20 Excerpted from the *Encyclopedia of the nations*. http://www.nationsencyclopedia.com/Europe/Finland-Government.html, 3. Accessed on May 31, 2015.
21 The Parliament of Finland, *Democracy's Long Road*, p. 18.
22 The Parliament of Finland. 2011. *Parliament's buildings and art*. Helsinki: The Parliament of Finland, p. 12.

23 The Parliament of Finland. 2000. *The voice of the people: Past, present, and the future*. Helsinki, Finland: The Parliament, p. 176.
24 The Parliament of Finland. 2011. *Parliament's buildings and art*, p. 11.
25 Tiihonen, S. 2008. The public sector in the conflicting pressures of democracy and efficiency. In *Very Many Democracies*, ed. P. Tiihonen, 53–85. Helsinki, Finland: Committee of the Future, Parliament of Finland, p. 66 (April).
26 Ibid. p. 84.
27 Suurla, R., M. Markkula, O. Mustajarvi .2002. *Developing and implementing knowledge management in the Parliament of Finland*. Helsinki, Finland: Parliament of Finland, Committee for the Future, pp. 2–3.
28 Ibid. p. 124.
29 Ibid. p. 113.
30 Ibid.
31 Excerpted from Suurla, R., M. Markkula, O. Mustajarvi. *Developing and implementing knowledge management in the Parliament of Finland*, pp. 114–121.
32 Suurla, R., M. Markkula, O. Mustajarvi. *Developing and implementing knowledge management in the Parliament of Finland*, pp. 124–125.
33 Adapted from Suurla, R., M. Markkula, O. Mustajarvi. *Developing and implementing knowledge management in the Parliament of Finland*, Picture pp. 41, 147.

3 Technology and Development

Public-Private Partnership

The emergence of Finland as a technology-driven economy has an evolutionary history: first by virtue of its geographical location and second because its political development as a country was nurtured by its pre-independent relationships with Sweden and Russia. The abundance of coniferous forests in the country structured its early development in the late nineteenth century as an agrarian society wholly dependent on wood and forestry products with respect to export commodities and local manufacturing activities. A combination of locally available raw materials and the rise of complementary manufacturing industries particularly in paper processing created a transformative economy that increased the value-added quality of Finland's primary agricultural exports. It also led to advancements in manufacturing technology, such as the development of paper machines as well as a redesign in the data acquisition systems used in the production processes.

3.1 Nature's Harvest and Industrial Processing

Although Finland relied heavily on the forestry industry until the 1950s, it imported most of its paper machines.[1] This situation changed when it exported the first paper machines while paying war reparations to the Soviet Union following WW II. By 1990, Finland became the world leader in paper machine industry and paper manufacturing technology, thus turning fate and adversity into an industrial success story that would test the enduring limits of organized capitalism and economic stability for years to come. As Nykanen and Paulapuro[2] point out, although the history of the Finnish paper machine industry can be linked to advances in production engineering and the design expertise of machine shops, a complex network of causes and effects can also be seen to underlie its history. The availability of cheap energy near the production area facilitated the expansion of the metal industry; the lack of free investment capital hampered the reform of industrial structures; corporate leadership culture was emerging and the education system encouraged studies in technical subjects and the acquisition of manual skills.

It is therefore worth the effort to illustrate that while new technology did not create enough intrinsic value to spearhead progress, the growing paper industry and the paper machine manufacturing that developed in its wake were dependent on a number of factors. As necessity became the mother of invention, the state stepped in to further create a horizontal alignment integrating the inchoate technological advancements in the manufacturing industry with curricula changes in the educational system, especially in pulp and paper manufacturing technology. Although there was no special education for paper machine engineers anywhere in the world in the early 1960s, students of mechanical engineering at the Helsinki University of Technology had long had the opportunity to choose paper technology as a major and graduate as mechanical engineers in the paper industry.[3]

As university education expanded to other regions of the country and educational responsibilities were divided, the University of Oulu was allocated machine design and chemical industry engineering, because these were expected to link suitably and to develop closer cooperation with the local pulp and paper machine industry. The existing professorship in chemical industry engineering at the university was changed into a professorship for paper industry engineering in 1971; in 1996, the Jyväskylä Technical Institute introduced a paper industry engineering course of study in its program; at the same time, the Tampere University of Technology established a professorship in paper industry automation.[4] By linking research outputs to industrial processes, key technology development programs executed by public research institutes became the major thrust to push industrial technology in a rising catch-up economy such as Finland. "By transferring their research and development (R & D) results directly or through spin-off mechanisms, public research institutes became one of the most effective sources of technology transfer."[5] Because Finland embraced the policy of cooperation among government, research institutes, universities and industry very early in its industrial development,[6] it made technology transfer and innovation critical elements in the successes and consequent economies of scale witnessed in the technology and manufacturing sectors.

3.1.1 Macroeconomic Adjustments and the Dirigiste State

Historical experience points to a simple truism that there is no absolute free-market system in the practical sense of the term. The philosophical debate between Keynesianism and classical laissez-faire economics suggests that both ideals have been complementary as opposed to being mutually exclusive. While Keynesianism focuses on the demand side of the macroeconomic spectrum as a necessity for growth and development, classical laissez-faire economics view growth as a function of supply and independent market adjustments that result from consumer choices in response to supply and demand. Hence, the sensitivity of the market to dramatic shifts away from equilibrium generally create automatic adjustments that drive price, supply and demand back toward equilibrium.

But where such is not the case, we are told to put the onus on the regulatory and restrictive policies of government that some suggest would stifle economic incentives for investment in the private sector—the main engine for capitalist economic growth. While "the role of government has traditionally been defined with *market failures*; its interventions have been targeted specifically to overcome the failures of the market economy to take sufficient care of important social tasks such as education, public safety, and infrastructure."[7] Furthermore, "with complex and dynamic policy problems that involve numerous private, civil, and public sector actors, systemic coordination is unlikely to take place entirely through the market mechanism."[8] In discussing the case of Finland, it is therefore worthwhile to position it in a comparative geopolitical context as it relates to other countries' economic experiences, as well as government's responses to them.

During much of the Great Depression in the United States, which began on October 29, 1920, the free-market system was not able to singlehandedly make the necessary adjustments needed to bring the economy back to a state of equilibrium or anything remotely like that. The American stock market, which had been performing well for almost a decade, crashed, plunging the country into its most severe economic downturn yet. Speculators lost their money, banks failed, the nation's money supply diminished, an estimated $30 billion dollars in stock holdings evaporated and companies went bankrupt and began to fire their workers in droves.[9] It did not end there as worldwide GDP fell by 15%, while "personal wealth, tax revenues, profit and prices dropped and international trade fell by more than 50%."[10] "Unemployment in the United States rose by 25% and in some countries by as high as 35%."[11]

In response to the economic crisis, President Franklin D. Roosevelt, who took office in 1933, acted swiftly to stabilize the economy and provide jobs and relief to the suffering citizens. Over the next eight years, the government instituted a series of experimental projects and programs that aimed to restore some measure of dignity and prosperity to many Americans.[12] This culminated in his New Deal program (1933–1938), which saw the creation of a medley of growth and employment programs that included the Emergency Banking Act, the Agricultural Adjustment Act, the National Industry Recovery Act, the Home Owners' Loan Act, the Tennessee Valley Authority, the Federal Deposit Insurance Act, the Federal Housing Authority, the Wagner Act and the Social Security Act. "From 1933 until 1941, President Roosevelt's programs did more than adjust interest rates, grant farm subsidies, or create short-term back-to-work programs,"[13] they also created new political coalitions that lent further support to his recovery initiatives, hence their broader receptivity within the American electorate. It was presidential activism yet unparalleled in US government administration, but one that ran counter to the core tenets of classical free-market liberalism. The Japanese bombing of Pearl Harbor on December 7, 1941, forced the United States' entry into WW II, but ironically, the war effort and related expenditures helped to stimulate the American economy and speed up the recovery effort that eventually ended the Great Depression.

Courting the Indigenous Market

Still, many instances of state dirigisme abound in the formative or later stages of the economic development of most countries. These actions range from deliberate interventionist measures such as direct grants of liberal loans to domestic businesses, quota systems and manipulation of existing tariff structures or by the institution of meandering bureaucratic bottlenecks as a way of protecting domestic industries from external competition. Dirigisme was introduced in France following WW II to promote industrialization and protect against foreign competition, and it was subsequently mimicked in East Asia. It took the form of indicative planning, which entailed government credit policies and subsidies, developing new technologies and the regulation of employment overseen by a special planning commission, the *Commissariat au Plan.*[14] It built an elite university for public administration, the *Ecole National d'Administration*, to train future state planners. The government also embarked on ambitious projects, encouraging the formation of national champions in large industry groups, such as the transportation system. In their policies toward the Stability and Growth Pact (SGP), which is part of the European economic governance system, the French government successfully influenced the reshaping of the fiscal policy architecture, loosening domestic economic constraints, and thus introduced a more dirigiste interventionism in the interpretation and implementation of the Pact.[15]

Just as France had done, "state authorities in Japan also pursued dirigiste or neo-mercantilist policies prioritizing selected sectors for rapid development and recruiting technocrats from the nation's elite schools for positions as planners in the state administration."[16] Through the organs of the Ministry of Trade and Industry (MITI), the country was able to offer liberal loans and engage in several protectionist measures to protect domestic manufacturing; by reinforcing the corporate culture of *keiretsu* (economic group), it ensured that domestic companies can pull their resources together to create a controlling influence and dominance in a market segment or industrial sector. This also enabled each major *keiretsu* to be capable of controlling nearly every step of the economic chain in a variety of industrial, resource and service sectors including banks and other financial and lending institutions. In like manner, South Korea sought to promote its own national economic group, the *chaebol*—providing long-term subsidized credit to various industrial groups, as well as putting up artificial export barriers to protect domestic industries. To expose an emerging economy to the uncertainty and vagaries of the free-market system so early in its development portends a very dangerous game. While countries should maintain control of the method and timing of their entry into the international market regime, they should also make sure that the domestic economy has achieved a sufficient degree of economies of scale or comparative advantage in critical sectors to be able to withstand the weight of exposure and competition in the global economy. It should also take note to develop enough slack or excess capacity in an area of specialization that could serve as a cushion during unforeseen economic hard times.

As China upgrades its hitherto command-and-control economy to become a major player in the international political economy, it has also learned to embrace dirigiste practices to its advantage. It forces countries to become heavily indebted to it, so that it can use that as subtle leverage to maintain de facto control over the debtor country's economic policy and often its politics. It also manipulates its currency by artificially devaluating it relative to the dollar, thus making her export goods cheaper in both international and domestic markets. This induces an artificial rise in the value of a comparative currency relative to the *yuan*, using the consequent high price of competing export products as a surrogate for protectionism and as an artificial barrier limiting how much of it is sold or purchased in China. Using the power of its Central Bank (The People's Bank of China), its monetary policy is often meant to prop up its slowing economy through a mix of targeted but unconventional liquidity and lending facilities.[17] "Since 1978, China has shown both flexibility and unyielding resolve in its continued pursuit of wealth and power; but she also knows that there are now things she wants from far beyond her borders—ideas as well as markets, raw materials and investments"[18]—and so must integrate strategically into the mainstream of international capital if the ideal of state dirigisme and market dominance is to succeed and endure.

For China, such integration has not come without political costs and questions raised about its commitment to a free-market economy, even if on a de facto level. The summer of 2015 saw a free-fall of China's stock market and the sudden decapitation of the hitherto earning capacity of its investment shares and stocks. As a series of investments portfolios drained within a matter of days, potential investors balked, and those who already had much committed in China's fledgling capital market began to seek the next exit out. Seeing what portends for the economy and the political hegemony of the Communist party, Chinese government officials led by the People's Bank of China "cut interest rates, increased the amount of money banks can lend out and pumped funds into the stock market—measures meant to boost domestic demand."[19] Also by artificially devaluing the Chinese currency (*yuan*), "authorities turned into a controversial growth-boosting tactic"[20] to the effect that a devalued *yuan* would make Chinese products cheaper and more competitive in the export market, thus giving an unfair advantage to Chinese export products, particularly when placed against American products.

For China, currency manipulation has always been a favorable macroeconomic policy tool as well as an adjustment mechanism for propping up Chinese exports. By arbitrarily pumping money into the economy, the People's Bank of China ensures that there will be enough liquidity in the economy to sustain business and private borrowing at an interest rate that does not dry up productivity or increase unemployment. This kind of loose-money policy would encourage growth, stabilize employment and make Chinese products artificially more competitive in the global market, to the chagrin of other competitors like the United States, South Korea and Japan. In the usual diplomatic parlance, Chinese government officials deflect evident criticisms by arguing that its actions were consistent with its objective of seeking a more market-determined exchange rate regime that would more properly reposition China as it integrates into the free-market system.

The same kind of state-led economic activism can equally be seen in a resurgent Russia under Vladimir Putin. Amid a slew of wrenching economic sanctions from the West against Russia following its foreign policy forays into other country's internal or territorial affairs (particularly Ukraine), Russia embarked on a series of measures to protect its domestic oil-driven export economy and soften the ever-growing pains of economic stagnation. This situation has been magnified by the plunge in oil prices from an average of almost $110 a barrel in the first half of 2014 to below $60 dollars a barrel in July 2015. While more than two-thirds of the country's export economy comes from energy, the consequent decline in the value of the *rouble* relative to the dollar and the rising indebtedness of several Russian businesses caused the country to sputter on the precipice of economic collapse.

In response, Vladimir Putin's push-back strategy involved a series of dirigiste inward-looking options set to withstand the economic ripple effects of the sanctions. "While the falling *rouble* made some export industries like farming more competitive, Putin's other import-blocking counter-sanctions against the West has allowed Russia to maintain a sizeable trade surplus"[21] as well as domestic industrial productivity to a level where the unemployment rate and industrial productivity could still be manageable at least in the short term. But to the extent that "Russian defences could still be tested by any one of a succession of possibilities—a continuous fall in oil price, a failed debt rescheduling plan by Russian firms that could drive them to imminent bankruptcy, and the possibility of more Western sanctions,"[22] there is little to say that other extra-free market approaches could not be put in place to stem further economic crisis. This situation is made even more profound by the simple fact that "when economies are on an unsustainable course, international finance often acts as a fast-forward button, pushing countries over the edge more quickly than politicians or investors expect."[23]

3.2 Finnish Cooperative Liberalism and Innovation System

The case of Finland, though different from that of Russia, China, South Korea, or the United States, nonetheless underscores a similar refrain regarding the centrality of state dirigisme in macroeconomic policy or in making critical adjustments necessary to contain decline and stagnation. Furthermore, it reinforces an enduring role for collectively organized economic actors in labor market and technology policy; in such a way that in Finland, collaborative technology policies and scientific research actively facilitated the rapid redistribution of resources to an emerging information, communications and technology (ICT) industry.[24] Instead of relying primarily on the reallocation of resource inputs through the free-market system, "institutionalized commitments between firms, banks, employer associations, trade unions, and state agencies created public goods on the supply-side of the economic spectrum."[25] The consequent growth in employment generated enough disposable income to create the conditions necessary for economic stability by striking a balance between production and consumption.

3.2.1 Statecraft and Political Will in Technology Policy

It can be said that the establishment of the Finnish Scientific Society in 1838 was a watershed moment in the evolution of the country's technology and science policy. In 1919, the tasks of the Society were transferred to the National Science Board, which had acted as a board of experts for the state's science-related questions. The board consisted of members of the Scientific Society and the

Finnish Academy of Science and Letters, which was formed in 1908. In 1948, the Academy of Finland was established, followed in 1950 by the formation of state research councils: the Research Council for the Humanities, and the Research Council for the Natural Sciences. "The state research councils were intended as a link between the state authorities and the research community; with the corresponding objective that it would afford the state the ability to direct more research funding and grants in support of the sciences."[26] Propitiously, this policy found a proactive political ally in the person of Urho Kekkonen (Agrarian Party) who had been both Prime Minister (1950–1953) and President of Finland (1956–1982) and whose personal and partisan leadership drove the trajectory of the science policy in its formative years.

The foundation for contemporary Finnish technology policy was laid after WW II in the early 1960s and 1970s but saw its more profound manifestation in the 1990s following the economic recession and decline in export trade caused partly by the dissolution of its major trading partner the Soviet Union in 1991. With Nokia transitioning from a manufacturer of rubber, paper, wood and forestry products into a major telecommunications provider, Finland suddenly became the leading country in the use of computers and mobile phone density. New communications technology products, service concepts and strong growth in international demand ushered in a competitive knowledge-based "new economy" that exalted a pioneering tradition that has typically defined Finnish culture. In many ways, it could be said that this was an opportune moment for Finland—a period in which the cumulative effects of sustained investments in science and technology, education and technical competence became more profound as drivers of economic and social change.

It took a proactive role on the part of the political parties to agree on the decision making needed to establish specific guidelines for the ensuing technology policy. As a means of strengthening the competitiveness of the national economy, independent decisions were made with regard to technology, social, industrial and science policies as well as the societal benefits that might accrue from them. Technology policy thus was aimed at building up a national research system and generating new knowledge and applications that address the needs of industry as well as society in general—a central tenet of an emerging Nordic-styled welfare state.

The path to realization has often been rocky, and at other times quite profitable. Heavy investment in existing industries and state-owned enterprises supported growth during the postwar era but hindered adjustment to a different economic environment when the bubble burst.[27] Following the economic recession of 1990–1993, Finland embarked on a dramatic restructuring of its economy from a paper producer to one of technology exporter. This became part of a broader shift to innovation-driven adjustment as reflected in spending on R & D, which increased from 1.2% of GDP in 1982 to 3.2% by the end of 1999.[28] The country simultaneously ranked higher than other EU countries in terms of per capita patent applications, innovation expenditure and high-tech value-added manufacturing.[29] Today,[30] over 60% of Finland's population now has cell phones, the highest proportion in the world and about twice the US proportion. It was the first country to have more wireless-phone subscribers than land-line subscribers and already had one of the highest teledensities—land lines per 100 inhabitants—on earth. Several surveys show that the country has more Internet users per capita than any other country and more than 35% of its population is online.

Finland's rise as a leader in technology manufacturing began when its 'fragmented telecommunications market forced equipment suppliers to compete with foreign producers; at the same time that the creation of a common Nordic mobile telephone standard gave the country a head start in developing technologies that would be used in the now dominant GSM standard."[31] Through the development of a robust innovation capacity, Finland was able to marry scientific outputs from research and development to an economic model that emphasizes innovation and competitiveness. By developing global clusters of innovation and technology diffusion, it was able to develop a mutually reinforcing technology model that drew from linkages and operations on a global scale. This objective was pursued chiefly through the promotion of research and development by national agencies like the State Technical Research Center (VTT) and the National Technology Agency (Tekes), which was formed in 1983.

This development saw the initiation of large-scale national technology programs and a dramatic increase in the overall level of research and development expenditures, surpassing that of many OECD countries. While an increasing part of the funding came from private corporations, the objective gradually moved from basic research toward more applied technological research and development activities.[32] Although the objective[33] of TEKES is to provide funding and expert services for R & D projects and promote the competitiveness of Finnish industry and the service sector, its activities were also meant to diversify and increase production structures, increase exports and create an enduring foundation for employment and social well being. Because it also sought to promote national and international networking, foreign companies conducting R & D activities in Finland are also welcome as TEKES customers. Selective funding has been the basis for Tekes's support. "Over the past 20 years, it has participated in and contributed to the creation of over 1000 innovations, hundreds of new companies, the creation of over 200,000 new jobs directly or indirectly, and the increased prosperity and social well-being in Finland."[34] On a more practical level, investments in science and the development of research activities that serve the needs of the business sector were investments for the benefit of economic growth. Hence, the rationale for accelerating research for industrial development was to generate new innovations through increased cooperation between the research institutes and businesses.[35] By working in "close partnership with the Academy of Finland, Sitra, Finnvera, Finpro, regional TE Centers for Employment and Economic Development, Tekes is able to offer comprehensive and technical support to any of its customers."[36]

Based on 2012 numbers, Figure 3.1 indicates a total government funding commitment to various research institutes and universities to the tune of €2.065 billion, with the Academy of Finland and Tekes receiving €320.7 million and €552.4 million, respectively. The largest sector for R & D funding went to the various research universities in the amount of €583.3 million, thus making up 29% of the total outlays for

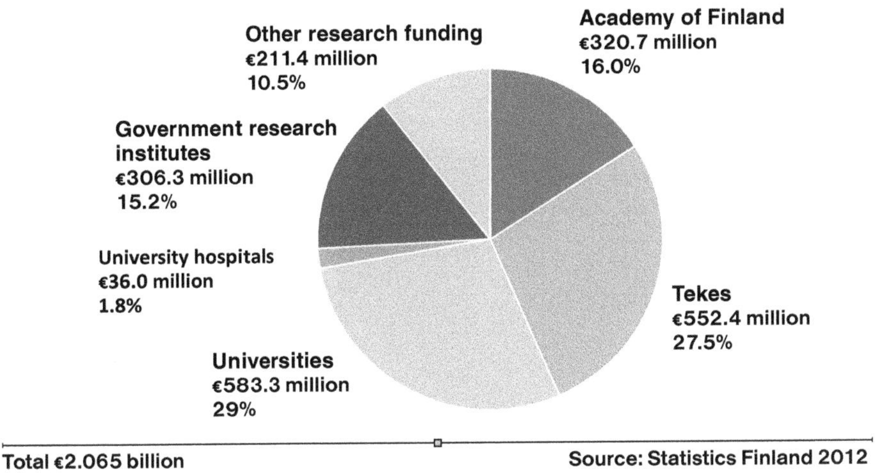

Figure 3.1 Total Funding and Institutional Outlays for Research and Development, 2012.

research and development. A unique feature of this funding program is the unwritten expectation that the various funding benefactors should collaborate with each other on their research findings and also that the findings be translated to industries where they could be converted to products that improve the quality of life as well as those that may have export value and potential. On the general premise that every problem has a potential solution, the purpose of research and development is to create new innovations that make seemingly difficult solutions more readily attainable. The Academy of Finland is one of the institutions that have over the years taken this mantra almost to its logical conclusion.

The Academy of Finland

Formed in 1948, the Academy of Finland has remained one of the main sources of research and development funding, leading to many creative and enduring scientific innovations. As an arm of the Ministry of Education, Science and Culture, it serves the primary role as Finland's leading source of R & D funding and is an active and major participant in Finnish innovative systems and scientific breakthroughs, as well as a strong voice in influencing science policy worldwide. The Academy's Board is made up of four functionally specialized research councils, each focusing on a specific area of human and scientific need: Research Council for Biosciences and Environment, Research Council for Culture and Society, Research Council for Natural Sciences and Engineering and Research Council for Health. Collectively they initiate clusters of research programs that focus on defined subject areas dealing with various social and scientific problems.

Figure 3.2 provides an overview of the funding distribution and opportunities provided by the Academy in 2011. This also includes funding for a various Centers of Excellence (CoE) across the country. Since 2000–2016, the Academy

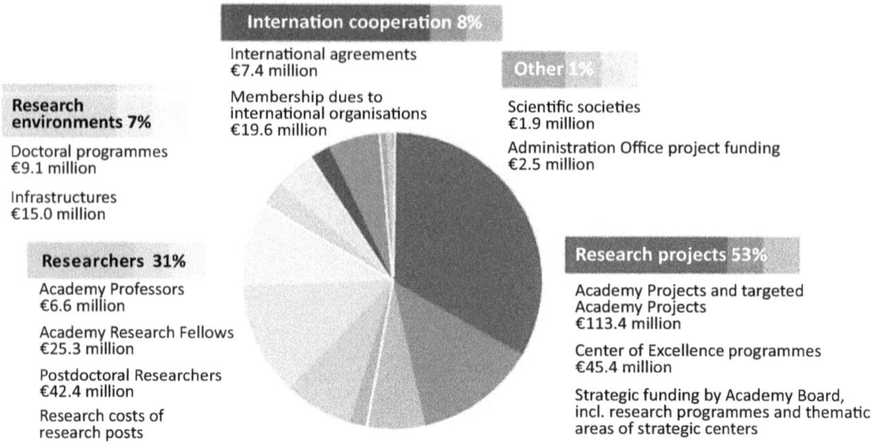

Figure 3.2 Academy of Finland: Distribution of Funding Opportunities and Layout 2011.

has funded approximately 98 Centers of Excellence with the primary objectives of improving the quality of Finnish research and international competitiveness, increasing its visibility and esteem and developing top-level innovative and research training environments. Among the research programs funded in 2012 were those that deal with climate change, computational science, nutrition and food, phonics and imaging technologies, sustainable energy and future living and housing.

The Academy also offers funding support for international research and scientific collaborations. It collaborates with the United States' National Science Foundation (NSF) through its Nordic Research Opportunity for NSF Graduate Research Fellows, and also the Virtual Center or *SAVI* (Science across Virtual Institutes). In 2011, it established *WiFiUS* (Wireless innovation between Finland and US), which also saw the NSF working in partnership with Finland's Tekes.

SITRA: The Finnish Innovation Fund

SITRA was created in 1967 as an independent public fund with an endowment capital of approximately €650 million, operated under the direct supervision of the Finnish Parliament. The idea was born out of the views of the then Governor and Chairman of the Bank of Finland Klaus Waris, as a way of celebrating Finland's 50th anniversary of its independence. The establishment of SITRA, otherwise known as the Foundation of Finnish National Fund for Research and Development, was seen as crucial for both accelerating the economic growth of Finland and the improvement of national competitiveness.[37] Initially, it provided R & D funding for the private sector from 1967 to 1983 before that mission was eventually transferred to Tekes.

In light of the evident changes in the global system, adjustments to globalization, institutional atrophy, disruptive technologies, stagnant economies, hunger and the problem of ageing societies, SITRA's role and strategy have evolved over the years in line with the needs of Finland. Its focus has shifted to one of seeking answers and solutions to the sustainable well-being of current and future generations, including economic, social and ecological sustainability. Under SITRA, Finland hopes to play a foremost leadership role in global sustainability development, gain new competitive advantages and create novel solutions that address the multiple problems of society. Its mission therefore is to provide answers to four vexing but enduring questions: How could we redefine well-being? How might we rethink sustainable societies? How might we reinvent the economy? How could we redesign the government? Figure 3.3 delineates how SITRA's role has progressed over the years from that of a technology developer (1967) to a venture capitalist and public sector shareholder in the 1990s and to a societal change agent beginning in the year 2000.

In the current context, SITRA's activities involve developing adaptive mechanisms for sustainable well-being, better service delivery systems, development of new business models and ecosystems, promoting new strategic thinking, healthy living and lifestyles, renewable energy, green economies, industrial symbioses and efficient utilization of resource streams. While SITRA's new focus areas are not about refurbishing old structures, they are more about starting from a clean slate anchored on the belief that we can achieve radical solutions through technology and its applications. The two areas that feature prominently in its current repertoire relate to ecological sustainability and the health management of individuals. In this regard, it has created a health information platform called *Taltioni*

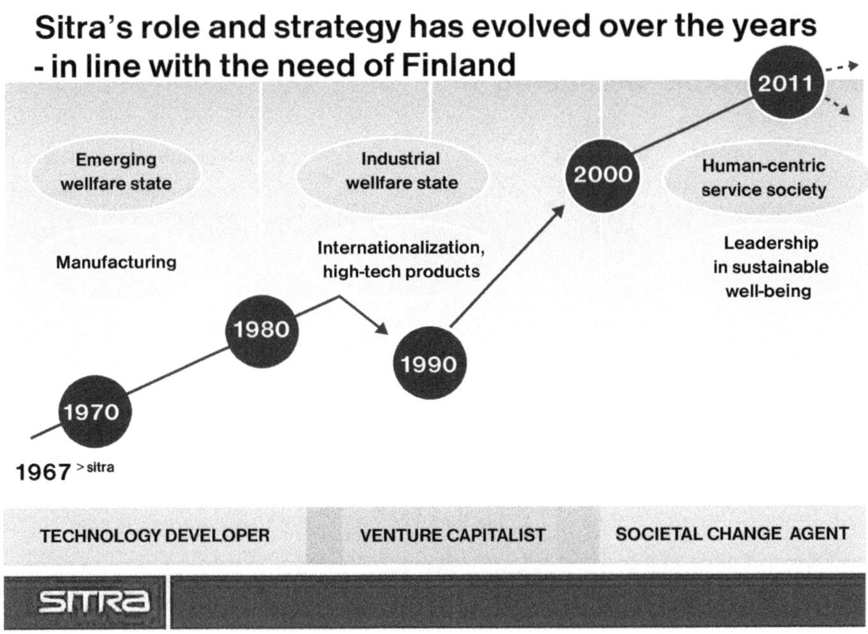

Figure 3.3 SITRA—From Manufacturing to Sustainable Well-Being.

"My record"—to which individuals can log in and save or retrieve their own health record and information. The idea behind this is that when people understand their own heath information and are able to monitor themselves accordingly, they can also take proactive action regarding their own personal well-being. They are less likely to get sick. If people are able to understand their own genetic status and its restrictions, they can also be able to know how to better take care of themselves in a complementary way and as a form of preventive medicine.

Also SITRA's new vision embraces the Nordic model of social justice—a proactive role for the government to reinforce equality and reduce all social disadvantages as much as possible. As Figure 3.4 illustrates, the model is built around three key objectives: economic sustainability, ecological sustainability and social sustainability. There is the belief that sustainable well being can only be achieved by a recognition of the symbiotic relationship between the environment, economy and society. Out of increasing concern for the dramatic social changes brought about by industrialization, urbanization, crime, unemployment and drug addiction in major urban cities, steps would need to be taken by the government in order to sustain social life for the future. In Finland, the problem of youth falling out of the mainstream of socially acceptable behavior has been on the rise over the years. They have resorted to utilizing the Netherland model known as "flag." This process allows authorities to monitor changes in youth behavior, academic performance and criminal tendencies as early as possible so that interdiction and remedial steps can be taken to avoid more serious and deleterious outcomes. The "flag' is thus a socially acceptable early warning system that can also save society a lot of headache and cost down the road.

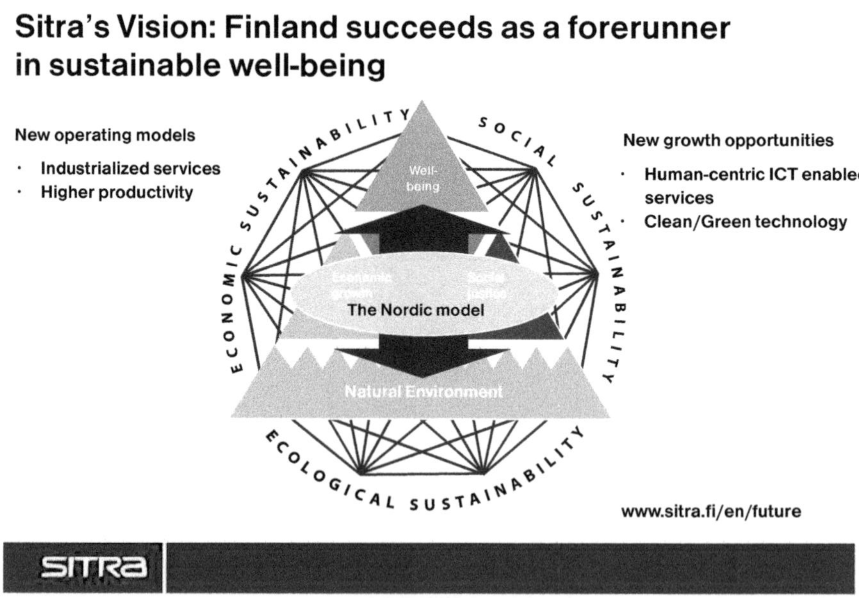

Figure 3.4 The Nordic Model for Sustainable Well-Being.

SITRA is also involved in developing what is known as *resource wisdom*. In light of the fact that there is so much food waste in most advanced societies, the agency is experimenting with the idea of collecting unsold restaurant and grocery meals, having them reprocessed and given out free of charge as street food to the hungry and needy. While this offers a way to save scarce resources, it also reinforces the twin virtues of social sustainability and social justice. One of the less-talked about consequences of industrial and economic development is that they can bring about dramatic changes in social structure and organization. The task of the welfare state thus is to find ways of accommodating or adapting to these changes in such a way that while not completely undermining the market, the role of the state in addressing fundamental anomalies that derive from it can be recognized.

3.2.2 Priming the New Economy through Technology Policy

The issue of how Finland was able to leap frog from a traditional agricultural and forest products-based economy to a high-tech and information society marvel within a short period of time has raised both interest and curiosity. While this feat has amazed a lot of people and state systems, it remained a perennial question that has attracted a series of explanations. It has also raised issues as to how much of Finland's successes can be transferred or replicated in other geopolitical and macroeconomic contexts. The reference to a 'new economy' signifies a break from the previous orthodoxy to something much narrower yet much more important economically—a permanent increase in economic growth driven by the productivity enhancing aspects of information technology and automation.[38]

Finland's adjustment following the economic recession of the early 1990s was characterized by the emergence of new industries as well as the improved cost competitiveness of traditional enterprises. Amid a growing concern on the need to shift rather than merely strengthen the country's comparative advantage, big business and industry adapted coordinating institutions to advance the course of economic restructuring.[39] As traditional tools used to manage adjustment in the past proved equally relevant in managing growth, Finland also achieved active centralization of collective wage bargaining and income policies that provided the *soft landing* needed for the restructuring to take hold.[40] While Finnish technology policy adopted a 'holistic' approach, whereby research was located within a broader national innovation system, it also advanced cooperation between public and private actors.[41] Because it was conceptualized as a broad, tripartite forum for managing adjustment, technology policy, in turn, facilitated cooperation with related initiatives in education and finance."[42]

Darius Ornston[43] provides a rather profound portrait of the origins, structure and spread effects of Finnish technology policy as reflected in the following accounts. While the earliest effort could be traced back to the 1960s, this policy took shape during the 1980s *in lieu* of concerns about Finland's reliance on paper exports and its bilateral trade relations with the Soviet Union. Following the agenda set by the Technology Committee (formed in1979), the government sought to increase R & D as a percentage of GDP from 1.2% in 1982 to 2.2 % by 1992. The creation of Tekes in 1983 supported corporate research and development projects, providing partial funding such that public funding not only replaced existing private sector efforts, it also viewed public and private research as complementary to each other. While public funding was central to Finnish technology policy, the institutional context within which it occurred was equally relevant.

The arrangements with the Technology Committee and Tekes helped to create a framework for managing adjustment during this difficult time. Because technological innovation was designed to engage clusters of firms, sectoral associations, research institutes and state agencies, it meant that "traditional patterns of inter-firm cooperation, ranging from sectoral associations to export cartels, were adapted to promote technological collaboration and diffusion among their members."[44] Taken together, the public R & D support and intense private sector collaboration facilitated the rise in research and development expenditure that made Finland the second most research-intensive nation in the European Union and a leader in innovation systems.[45] Technology policies empowered R & D directors much more than management, linked firms to a growing network of public and private sector actors and contributed to a broader attitudinal change among Finnish employers.[46] While it created a new role for traditional actors such as the Ministry of Trade and Industry, it also did so without directly suppressing wages, limiting employment stability or reducing non-wage costs.

As the low political costs associated with public funding of R & D became more pronounced, trade unions raised the costs associated with alternative reform strategies while supporting measures designed to improve technological competitiveness. There was a collective recognition that all sectors of the economy must sacrifice if the restructuring process was going to be successful. But to do that required a level of transparency and ability to communicate with all relevant stakeholders in a timely and effective manner. In the end, Finnish technology policy was a political, economic, scientific and cultural project. Seen as a product of liberal reforms in finance, competition policy, trade and taxation reforms, the country's "success relied mainly on a long-standing tradition of successful inter-firm collaboration in the private sector, a dense network of national employer associations, sectoral organizations, and a unique constellation of research and financial institutions."[47]

3.3 The Nokia Revolution: Pioneering Ethos

The economic crisis of the 1990s brought about a radical departure from the traditional economic model and product focus than was previously the case in Finland. In 1991,[48] the GNP fell below .06%, and unemployment rose from a low of 0.4% to a double-digit 10%. As the public debt balance turned into a crunching deficit; Finland's currency had to be devalued and was exposed to the floating exchange rate mechanism. These developments drove the Finnish banking sector into a severe fiscal and liquidity crisis, and the state once again had to save the bank's solvency with huge subsidies that, ironically, led to more public sector deficits. Something needed to be done, and fast. As a result of broad-based state-driven restructuring plan, "high technology products surged as a share of exports, surpassing even pulp and paper products"[49] and other traditional goods as the country moved swiftly to take a commanding lead in the emerging telecommunications and consumer electronics market. Enter Nokia to the rescue: a Finnish conglomerate whose traditional business focus has been on forestry and wood products, rubber, cable, electronics and power generation systems.

Nokia's revolutionary history[50] dates back to 1865, when then mining engineer Fredrik Idestam (1838–1916) set up his first wood pulp mill at the Tammerkoski Rapids in southwestern Finland. A few years later, he opened a second mill on the banks of the Nokianvirta river, inspiring him to name his company Nokia AB in 1871. It eventually became Nokia Corporation following a tripartite merger in 1967 of Idestam's Nokia AB, Finnish Rubber Works, and Finnish Cable Works. Regulatory and technological reforms and deregulation of the European telecommunications industries since the late 1980s helped to stimulate competition as well as customer demand that played a critical role in Nokia's later success. The mid-1990s saw the company divest itself from its traditional holdings and operations to focus on telecommunications, consumer electronics infrastructure and related services. "By the year 2000, it had outperformed such established players like Ericsson and Motorola to assume a leading 35% share of the mobile handset market."[51]

As the company became a central feature of the Finnish economy, it was also an indispensable engine that quickened the country's economic recovery, as well as a symbol to state authorities and the public that the national emphasis on innovation-driven macroeconomic adjustment was working. "The country heavily invested in Nokia with Tekes (The Finnish Funding Agency for technology and Innovation) financing 26% of the company's projects for much of the 1990s; in return, Nokia gave back to Finland with huge tax revenue and employment of highly skilled workers."[52] The increasing demand for high-skilled workers ultimately helped to drive changes in an educational system whose schools are now ranked the best in the world. At the same time, Nokia's economic ascendancy was transformative in the sense that it rekindled one of the positive ethos of Finnish culture that there is always an answer to every human project if we keep working at it. While the company relied on domestic resources and suppliers to manage an extraordinary increase in production,[53] its activities "constituted approximately 2.7% of GDP and 60% of the total value of the Helsinki stock exchange in 2002."[54] The economic spread of Nokia's success led to the creation of many jobs and spin-offs in the ICT sector and tertiary industries, supply chains and intermediate providers of parts and services (Figure 3.5). In addition, Finnish labor productivity growth averaged 2.4% per year from 1995–2002, a performance that outstripped the US's 2.1 percent by a significant margin.[55] Only Ireland and Greece scored higher margins than Finland during this period of time.

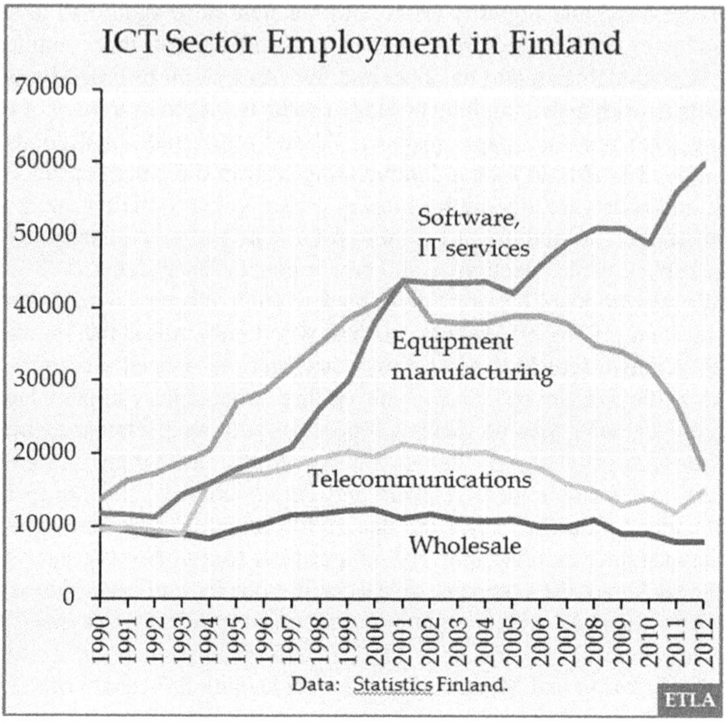

Figure 3.5 Growth and Decline in Employment in Finland's ICT Sector 1990–2012.

As Nokia continued to deliver surprising technological breakthroughs in innovation, its global market share soared from 10% to above 30% in a matter of years as the company bulldozed through tough markets like China and Brazil with astonishing ease.[56] Between 1995 and 2000, Finland enjoyed the lowest Internet access price and highest Internet penetration in the OECD countries;[57] but as the telecommunications and the mobile phone market became more saturated with increasing competition from major industry actors such as Samsung, Apple and Ericsson, the company (Nokia) that used to account for 20% of Finland's GDP saw its sales decline with a growing loss of revenue and a decline of about €200 billion in market value.

Megan McArdle explains Nokia's predicament quite succinctly:[58] Between 1998 and 2007, the company contributed a quarter of Finnish economic growth, its spending on research and development accounted for 30% of the country's total and it generated nearly a fifth of Finland's total exports. Nokia was paying as much as 23% of all Finnish corporate tax revenue. By 2013, "its percentage of Finland's GDP had fallen from 4 to 0.4%, its share of exports more than halved, and its share of Helsinki stock exchange market capital plummeted from 70 to 13%."[59] Amid declining sales, market share and investor confidence,

Nokia sold all of its devices and service businesses to Microsoft for a combined sum of €5.4 billion and refocused its core business strategy on three technology centers: Nokia Networks (network systems), HERE (location maps and directory) and Nokia Technologies (advanced technologies and innovation). With this sale and consequent reorganization, a new normal seeks to unravel Nokia's traditional corporate culture and Finnish identity. Little consolation that the "company that built a country seemed to have lost its aura and relinquished controls."[60]

Today, Nokia is still in the same throes of uncertainty as its partnership with Microsoft, which began in 2013, seems to be unravelling. Microsoft entered into the deal as a way of giving it a controlling presence in the smartphone and mobile handset business. In July 2015, Microsoft admitted that it was "writing down about 80% of the $9.4 billion deal for Nokia's handset business, and will cut 7,800 workers from its global payroll—mostly in its mobile phone operation. This comes a year after it had embarked on a restructuring plan that cut about 18,000 jobs"[61] In its wake have been left a lot of unanswered questions about the danger of a country's fortunes being so tied to the success and failure of a single company.[62] If Nokia goes, what happens to Finland? Why was the paradoxical nature of the high-tech environment so easily lost on Finnish political and economic leaders?

It should be a known fact that the technology industry operates in a highly uncertain and fluid environment, where new developments have a short life span and where continuous innovation or even wholesale transitions to brand new and exotic systems remain the norm for survivability. Product lines and new innovations have a very short time to maturation, and in an evolving market like this, competition is oftentimes most brutal. As was the case with Nokia, competitors like Apple, Google, and Samsung never let up, and by dictating the tune of consumer taste by pushing novel applications like the iPhone and iPad unto the market, they were able to set the trend as well as shape market demand. They gave the consumers what they *should* buy, but not necessarily what they *wanted* to buy. And it worked.

Nokia may have been caught off guard, but as it rested on its laurels and relied on its traditional product lines; it lost market share to the competition amid declining investor confidence. Another critical element in Nokia's evident decline may be an inherent durability bias in the technology sector. New innovations have a limited time span to shape evolving market trends, and if and when they do, they are more likely to reach critical mass sooner than later. To base a country's long-term economic sustainability on a highly fluid and dynamic sector such as this suggest that policy makers and economic actors' optimism was overshadowed by existing short-term benefits that, invariably, did not take into account the probability of future uncertainties and long-term economic inertia. In the end, there is only so far that Finland can go with technology and innovation and still remain Finnish in all its ramifications. This is the uncertain and foreboding challenge now and for the future.

Notes

1 Nykanen, P. and H. Paulapuro. 2005. *Around the roll: A century of Finnish paper machine and papermaking technology*. Helsinki, Finland: The Finnish Academies of Technology, p. 15.
2 Ibid. pp. 16–17.
3 Ibid. p. 179.
4 Ibid. pp. 179–180.
5 Cheng, W-P. 2001. *A strategic approach of building national technology strength: Finland as case study.* Master Thesis., Sun Yat-sen University, Taiwan, p. 16. Also see CRN: etd-0627101-121030.
6 Ibid. p. 17.
7 Hamalainen, T. 2013. *Building blocks for a new socioeconomic model*. Helsinki, Finland: SITRA, Version 10, p. 10.
8 Ibid. p. 10.
9 http://www.history.com/topics/new-deal, pp. 1–7. Accessed on July 3, 2015.
10 Frank, R. H. and B. S. Bernanke. 2007. *Principles of macroeconomics*, 3rd ed. Boston, MA: McGraw-Hill, p. 98.
11 Ibid.
12 http://www.history.com/topics/new-deal, pp. 1–7. Accessed on July 3, 2015.
13 Ibid. p. 6.
14 Encyclopedia Britannica. http://www.britannica.com/topic/dirigisme, p. 1. Accessed on July 3, 2015.
15 Clift, B. 2006. The new political economy of Dirigisme: French macroeconomic policy, unrepentant sinning and the stability and growth pact. *British Journal of Politics and International Relations* 8 (3): 388–409.
16 Encyclopedia Britannica. http://www.britannica.com/topic/dirigisme, p. 1. Accessed on July 3, 2015.
17 *Economist, The*. The People's Bank of China, November 22, 2014, p. 14.
18 *Economist, The*. What China wants, August 23, 2014, p. 48.
19 Hilsenrath, J. and B. Blackstone. 2015. Cheaper currency has global impact. *The Wall Street Journal*, August 12, p. A1.
20 Ibid.
21 *Economist, The*. 2014. Russia's wounded economy, November 22, 2014, p. 9.
22 Ibid.
23 Ibid.
24 Ornston, D. 2006. Reorganizing adjustment: Finland's emergence as high tech leader. *West European Politics*. 29 (4): 784–801, p. 784.
25 Ornston, D. Reorganizing adjustment: Finland's emergence as high tech leader, p. 785.
26 Murto, E., M. Niemela, and T. Laamanen. 2006. *Finnish technology policy from the 1960s to the present day: Public sector activities promoting research and development*. Helsinki, Finland: Ministry of Trade and Industry, p. 8.
27 Ornston, D. Reorganizing adjustment: Finland's emergence as high tech leader, p. 790.
28 Asplund, R. 2003. *Flexibility and competitiveness: Labour market flexibility, innovation, and organizational performance*. Helsinki, Finland: Taloustieto Oy, 10. Also see Ornston, D. *West European Politics*. 29 (4): 784–801, 788.
29 European Commission. 2001. European competitiveness report 2001. Luxembourg: The Office for Official Publications of the European Communities, 38. Also see Ornston, D. *West European Politics*. 29 (4): 784–801, 788.
30 Colvin, G. 1999. We see the future, and it's Finland. *Fortune* 24(4): 1–4, p. 1.
31 Rouvinen, P. and P. Yla-Antilla. 2003. Little Finland's transformation to a wireless giant. In *The global information technology report* 2003–2004, ed. S. Dutta, B. Lanvin, and F. Paua, pp. 87–109. New York: Oxford University Press.

32 Cheng, W-P. *A strategic approach of building national technology strength: Finland as case study*, p. 78.
33 Ibid. p. 80
34 Finnish Academies of Technology. 2004. *High technology Finland* 2004. Helsinki, Finland: Finnish Academies of Technology, p. 16.
35 Murto, E., M. Niemela, and T. Laamanen. *Finnish technology policy from the 1960s to the present day: Public sector activities promoting research and development*, p. 11.
36 Finnish Academies of Technology. *High technology Finland* 2004, p. 16.
37 Murto, E., M. Niemela, and T. Laamanen. *Finnish technology policy from the 1960s to the present day: Public sector activities promoting research and development*, p. 29.
38 Daveri, F. and O. Silva. 2004. Not only Nokia: What Finland tells us about the new economy. *Economic Policy* (April): 117–163, p. 119.
39 Ornston, D. Reorganizing adjustment: Finland's emergence as high tech leader, p. 792.
40 Ibid. p. 791.
41 Lundvall, B-A, ed. 1992. *National systems of innovation: Toward a theory of innovation and interactive learning*. London, UK: Pinter.
42 Ornston, D. Reorganizing adjustment: Finland's emergence as high tech leader, p. 792.
43 Ibid. pp. 792–794.
44 Schienstock, G. and T. Hamalainen. 2001. *Transformation of the Finnish innovation system: A network approach*. Helsinki, Finland: Finnish National Fund for Research and Development.
45 European Commission. 2001. *European competitiveness report* 2001. Luxembourg: The Office for Official Publications of the European Communities.
46 Ornston, D. Reorganizing adjustment: Finland's emergence as high tech leader, pp. 793–794.
47 Ibid. pp. 796–797.
48 Cheng, W-P. *A strategic approach of building national technology strength: Finland as case study*, p. 37.
49 Ornston, D. Reorganizing adjustment: Finland's emergence as high tech leader, p. 789.
50 Nokia. 2015. Our Story. http://company.nokia.com/en/about-us/our-company/our-story. Accessed on June 15, 2015, pp. 1–2.
51 Stenbock, D. 2000. *The Nokia revolution: The story of an extraordinary company that transformed an industry*. New York, NY: Amacom.
52 Kelly, G. 2013. Finland and Nokia: An affair to remember. *Business*, October 4, 2013. http://www.wired.co.uk/news/archive/2013-10/04/finland-and-nokia, p. 3. Accessed on June 12, 2105.
53 Moen, E. and K. Lilja. 2005. Change in coordinated market economies: The case of Nokia and Finland. In *Changing Capitalisms: Internationalization, Institutional Change and Systems of Economic Organization*, ed. G. Morgan, R. Whitley and E. Moen, 352–379. Oxford: Oxford University Press, p. 362.
54 Ornston, D. Reorganizing adjustment: Finland's emergence as high tech leader, p. 789.
55 Daveri, F. and O. Silva. Not only Nokia: What Finland tells us about the new economy. p. 124.
56 Kuittinen, T. 2013. A brief history of Nokia: When the future was Finnish. URL: http: brg.com/2013/09/03/nokia-history-analysis-microsoft-acquisition,September 3, 2013, p. 2. Accessed on June 15, 2015.
57 Daveri, F. and O. Silva. 2004. Not only Nokia: What Finland tells us about the new economy, p. 128.
58 McArdle, M. 2013. As goes Nokia, so goes Finland? *Bloomberg View*, September 3, 2013, p. 1–2. URL: http://www.bloombergview.com/articles/2013/-09-03/as-goes-nokia-so-goes-finland. Accessed on June 15, 2015.
59 Kelly, G. Finland and Nokia: An affair to remember. http://www.wired.co.uk/news/archive/2013-10/04/finland-and-nokia, p. 5. Accessed on June 12, 2015.

60 Ibid. p. 1.
61 Ovide, S. 2015. Microsoft's misdial cost$ 7.6 billion. *The Wall Street Journal*, Vol. CCLXVI (7), Thursday July 9, 2015, P. B1.
62 Powell, N. 2011. How Finland's economy became hooked on Nokia. *The Globe and Mail*. October 26, 2011, p. 1.

4 Economic Growth, Research, and Innovation Quality

Finland has a highly industrialized, free-market economy[1] with a per capita output equal to that of other western economies such as France, Germany, Sweden or the UK. The largest sector of the economy is services (65.5%), followed by manufacturing and refining (31.6%). Primary production is at 2.9%. The United States is Finland's third most important trading partner outside of Europe. With a 3.4% share of imports in 2009, the United States was Finland's ninth-largest supplier. Major exports from the United States to Finland continue to be machinery, telecommunications equipment and parts, metalliferous ores, road vehicles and transport equipment, computers, peripherals and software, electronic components, chemicals, medical equipment and some agricultural products. The main export items to the United States are electronics, machinery, ships and boats, paper and paperboard, refined petroleum products, telecommunications equipment and parts.

The Finnish economy[2] has made enormous strides since the severe recession of the early 1990s. Finland successfully joined the Euro Zone and outperformed euro-area partners in terms of economic growth and public finance. Exports of goods and services contribute over 37.4% of Finland's GDP. Metals and engineering (including electronics) and timber (including pulp and paper) are Finland's main industries. Following a period of sustained and robust growth, the Finnish economy suddenly slowed in the wake of the international financial crisis and stiff competition from other manufacturers of technology and mobile communications equipment. GDP growth shrank from 0.9% in 2008 to - 8% in 2009 (the sharpest contraction since Finland gained independence from Russia in 1917). Exports declined 32%, and unemployment climbed to 8.2%. Figure 4.1 denotes the devastating consequences of the economic depression of 1991 following the dissolution of the Soviet Union and the drying up of its market for Finnish exports. The downward spiral in employment continued until the modest recovery of 2009 and then plummeted again. While Finland's unemployment has maintained a single-digit rate from 1956 to 1991, it has rarely generated the type of spread effects one would except of most diversified economies. The unemployment experiences of the 1990s hit hard on the technology sector and related service industries simply because it was the largest employer of Finnish labor.

Therefore, the economic crisis could be seen as a sector-driven market adjustment that, invariably, spread its tentacles throughout the whole economy including

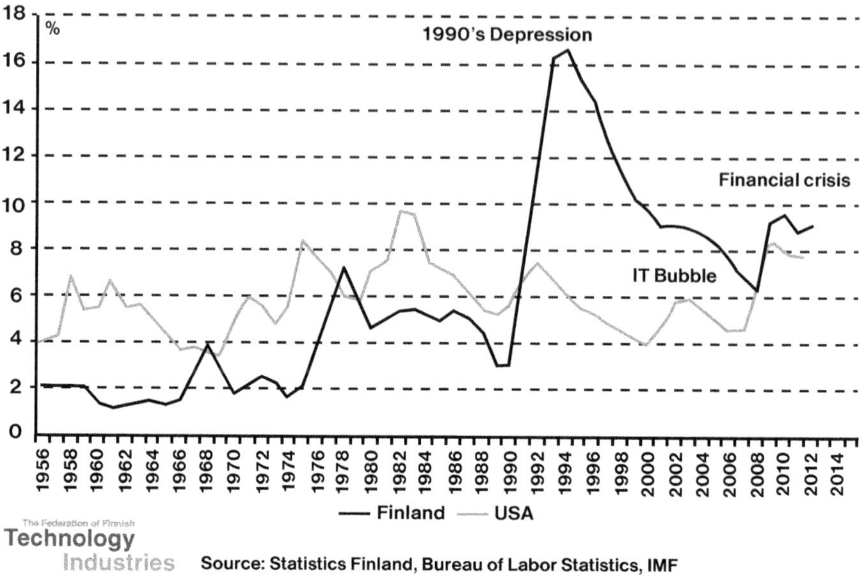

Figure 4.1 Finland and USA: Annual Unemployment Rate, 1956–2011.

the banking and financial sectors. As a result of the decline in the level of exports and the consequent expansion of the public sector in an effort to address competing issues of the welfare state, the national debt rose from a low of €60 billion in 2005 to well over €90 billion by 2013 (Figure 4.2). This confounded much of the recovery effort as well as the ability of the government to put in place crucial macro stabilization mechanisms for recovery and growth. While the above sets the broader premise from which to view Finland's economic growth and development over the years; but in light of the current economic realities and changes in the extant production structure, it should also suffice to point out that the technology industry has remained the most important driver of economic growth as well as decline.

The reasons are not far-fetched. As of 2013, the technology sector made up about 55% of total Finnish exports and comprised 80% of private sector research and development investments. With about 290,000 people employed directly in the sector and another 430,000 in related or tertiary industries, it made up of about 30% of the Finnish labor force. With rising production costs and generally declining growth prospects in industrialized economies, the technology sector faced further obstacles in terms of price competition, undiversified production structure and product portfolio and an aging population that could affect the supply of skilled labor. There may be other related but nuanced issues tied to the public sector financing model utilized by most Finnish firms and start-ups. "Grants, subsidies, and governmental loans without stringent repayment clauses require strong

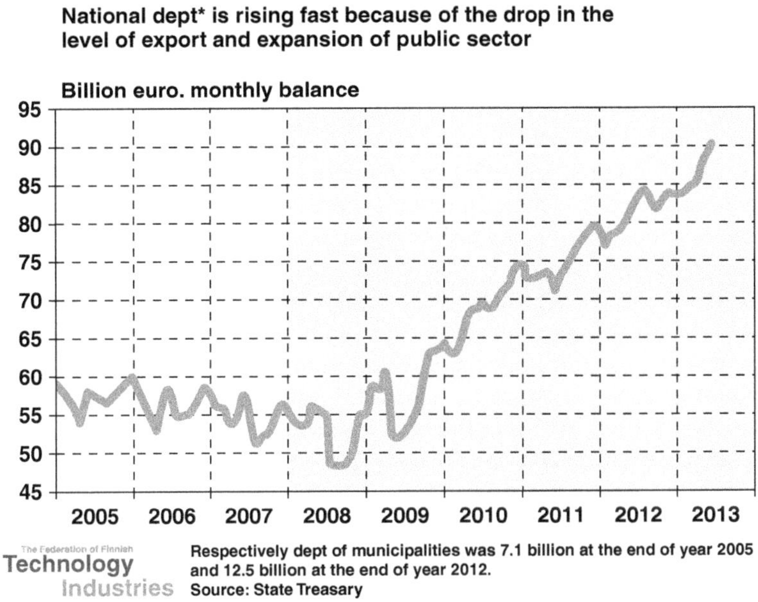

Figure 4.2 National Debt Profile, 2005–2013.

monitoring practices to avoid moral-hazard dilemmas because these instruments are the preferred funding sources in both market and non-market-oriented (public utility) companies."

It may, in fact, be the case that overly liberal loans granted to market-oriented firms by the government to speed up economic productivity and employment, without a clear assessment of long-term prospects for repayment, may have made potential default possible if not imminent. They induced companies to take on more ambitious projects when their resources and skills remain untested, thus exposing them to higher risks of failure. Paradoxically, "government organizations that specialize in venture capital financing are, by definition, not proper venture capital entrepreneurs because they invest taxpayer's resources and do not face the threat of operational default in the case of investment failure."[3] And neither are "the upside gains from successful investments reflected in the investment manager's personal wealth because civil servants in Finland do not receive performance-based compensation."[4] This raises the issue as to whether public sector investments are guided by the same strategic interest for profit making, or at best value creation, as you would have in the private sector.

4.1 Technology and the Economy

The revolution in communication technology has changed the way people exchange information and interact with each other. This factor has made it possible for Finland to become an intensive user of ICT devices and services but also a leading producer and pace setter in the industry. This includes such applications as the Internet, discussion boards, mobile broad band, e-services, mobile phones and many others. "While a total of 88 percent of all households have an Internet subscription; subscription for mobile broadband has increased by 12% over the years, with 43% of households subscribed to it."[5] The Internet is increasingly used for online communications, information searches, e-services, filing applications and making online payments. It has become an indispensable part of everyday life.

To serve the domestic need and a rising export demand for these products and services, numerous companies have made a name for themselves opening up new business opportunities and technologies in various fields of engineering, manufacturing, software development, operations management and consulting. By creating jobs that attract an increasingly skilled labor force, the rise in employment means that people will have more disposable income to spend in the purchase of goods and services. The multiplier effect of consumer activities and market forces leads to further increases in the demand for goods and service as well as economic productivity. Figure 4.3 delineates five industrial clusters that make up the technology industry in Finland: electronics and electro-technical, mechanical engineering, metals, information technology and consulting engineering. By 2012 calculations, the mechanical engineering sector generated more revenue (€27.9 billion) and more employment (131,000) than the electronics and IT sectors combined. Much of the productions in this category are capital-intensive fixed items and industrial tools that can last a

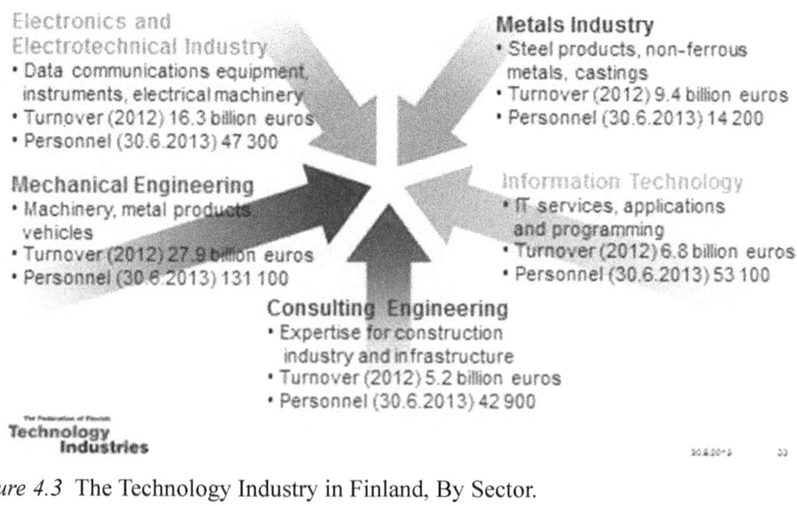

Figure 4.3 The Technology Industry in Finland, By Sector.

long time and hence may not be easily affected by changing market trends. On the other hand, electronics and information technology products are more regularly dictated by new technological applications, software and shifting consumer tastes. The ensuing uncertainty makes the market more volatile, hence more prone to the adverse effects of the highs and lows of conventional macroeconomic indicators like interest rates, unemployment, public indebtedness, competition and international crises.

It can also be argued that for the simple reason that much of Finland's exports from the technology industry have gone to neighboring European countries, it lacks the global reach critical for achieving international market dominance (Figure 4.4). With approximately 65% of the exports going to Western and Eastern Europe to the tune of €17.3 billion, the export prices would be affected by changes in the market value of the euro and changes in the economic fortunes of these countries, some of them still struggling to make durable transitions from the Soviet era command economy to a liberal market regime. Although export of technology-related services amounted to approximately €12 billion, suffice it to say that a marginal expansion in the North American and Asian markets would have created the necessary buffer and slack for the industry to weather any economic downturns or slide in demand from the existing European markets.

This may have been part of the calculation in the inauguration of the pilot program known as *FinChi—Gateway to China*, with the primary goal of transferring expertise between Finland and China and facilitating Chinese investments and businesses in Finland—a program that would enable China to use Finland as a

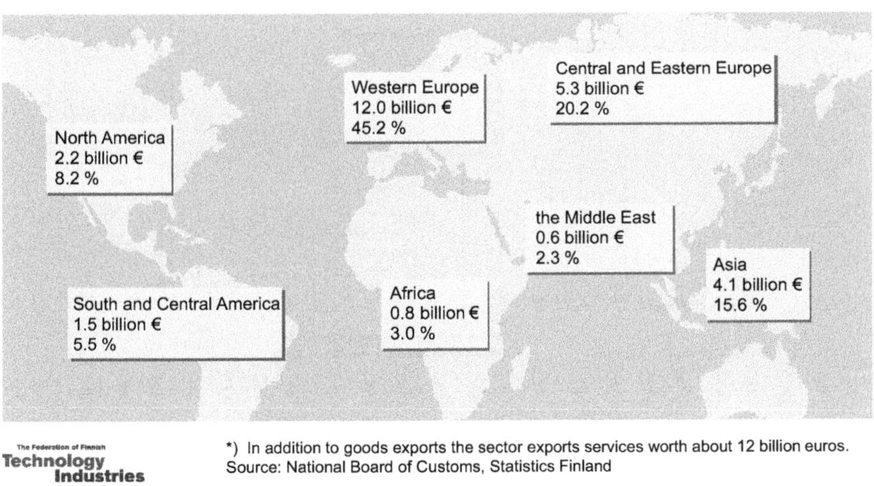

Figure 4.4 Technology Industry Export of Goods by Geographical Area and Market, 2012 (Total goods exports €26.5 billion).

bridgehead to other western markets.[6] As part of Teknia's international partnership strategy, *FinChi* is expected to provide major benefits for high-tech businesses in Finland and help eliminate barriers to doing business in Asia. The program also stands to benefit from Teknia's agreement with other international universities in the United Kingdom and the United States.

The evolution of Finland's economic growth and performance has essentially been tied to historical and political milestones in its emergence as a sovereign state (Figure 4.5). While most of these events have been externally induced, others have more or less been either coincidental or beyond the immediate control of the national leadership. Fluctuations in GDP growth were affected by the civil war (1917–1918), the Great Depression (1929–1939), WW II (1939–1945), post-WW II reparations to the Soviet Union (which ironically spurred economic growth in Finland), the 1973 and 1979 oil crises, the fall of the Soviet Union (1989–1991), Finland's entry into the Eurozone, and the global financial crisis of 2008.

The key issues raised by these historical events and their potential consequences for Finland relate to the scope of exposure and integration of the national economy to the international market regime and its potential immunity from adverse political and foreign policy developments in other countries or in the international system. In 2012, the *Global Enabling Trade Report* ranked Finland in 6th place while the United States was ranked 23rd out of a

Figure 4.5 Production Volume Development in Finland: Annual GDP Growth in Constant Prices, 1900–2015.

total of 132 countries. This assures that Finland's economic performance and growth would continue to be affected by global market or political forces in the foreseeable future. While much of this is unavoidable, the dilemma is that Finnish technology and industrial products need increasing international market exposure and integration as a way of sustaining the needed revenue base to grow the national economy and sustain its current standard of living and model welfare state.

Nonetheless, while it would be hard to determine where a crisis will arise and in what form and scope it will take, it would seem somewhat precarious to base one's economic survival on global events that one could not, more or less, dictate. Technological innovations can spur growth, but at some point they reach critical mass or what economists call point of diminishing marginal returns. Even when new innovations are brought forth to sustain growth, sooner or later, they too succumb to the high turnover rate of the technology industry and the inertial pull of diminishing marginal utility, as market trends become less predictable. As new innovation and breakthroughs are developed, they remain constantly under siege as other, perhaps more revolutionary, alternatives or replacements are discovered. In order to grow its economy and remain relatively solvent, Finland, though not exclusively, needs a stable international market regime as well as a peaceful world. But how much of that is guaranteed or predictable? This should be one of the overriding questions driving Finland's ongoing innovation policy.

4.2 Building Value Chain through Innovation Policy

The term *innovation* means finding new ways of doing things better, and this has been the key driver of Finnish technological development. "Innovation and thinking 'outside the box,' in terms of setting technological or business goals, underpinned much of the massive growth in telecommunications and mobile technology that spearheaded Finland's early rise in this industry."[7] "Thinking outside the box and taking risks, learning from your successes as well as failures, and always striving to go one step further"[8] are the hallmarks that make innovation possible. Finland has achieved a remarkable record as a high-technology country and a competitive economy over the last few decades, this is driven in part by its long-term emphasis on educational equality and public-private collaboration by industry, research institutes and universities. There is a healthy feeling that "public funding of technology and development works, and that the encouragement of public-private cooperation in this regard has been a major factor in the Finnish success story."[9]

Other equally critical factors include the relatively small size of Finland compared to its neighbors, as well as its pragmatic approach to problem solving; the fact that "it is often relatively easy for people to agree on things, and when the facts are put on the table they just roll up their sleeves and get on with the job."[10] By linking basic research more closely to successful business innovations, the country is able to develop a systematic and consistent framework for evaluating how science, technology and innovation effectiveness best work together and can be effectively leveraged to serve current and future human needs. Innovations in environmental and ecological management have led to discoveries that cover a broad range of areas that "promote more sustainable-minded construction, helping people use less water, cleaning up oil spills more effectively, and ensuring better indoor air quality."[11]

The area of health care and the life sciences has also seen very rapid development in Finland. "Basic research in enzymes and drug discovery has resulted in more efficient industrial processes and new approaches to functional foods, as well as more effective therapies for treating diseases and the development of new tools for faster and more accurate diagnoses."[12] At the forefront of this national endeavor are agencies[13] such as Tekes, VTT, Finpro and SITRA. Tekes (the Finnish Funding Agency for Technology and Innovation) has been the main publicly funded organization for financing applied and industrial R & D in Finland. Its primary objective is to promote the competiveness of industry and the service sector by enhancing the country's technological potential—in areas such as diversifying production structures, increasing production and exports and creating a more solid foundation for prosperity today and into the future. Programs focus on specific sectors of technology and industry and are designed to give business access to laboratory research results as well as results from international R & D cooperation.

The Technical Research Center of Finland (VTT) is the largest multidisciplinary research organization in Northern Europe, offering R & D, testing, product approval, certification and information and venturing services that help customers create new products and generate new businesses. VTT and its 2,700 professional employees possess cutting-edge expertise across a broad spectrum of technologies for creating globally competitive technological innovations. In 2006, it filed approximately 280 individual patents, and has over the years announced some 150 inventions annually. VTT's operations and services cover the entire innovation process, from ICT, functional materials, nano technology, bioenergy, enzymes, laser welding, near-field communication technology (SmartTouch project), new energy generation technology and biomaterials for tissue engineering. By building synergies with other research facilities and heavy involvement in EU projects and national innovation networks, VTT has been able to enable its customers see beyond the business they have today to the business they could have in the future.

Finpro's traditional role is to act as a partner for Finnish companies, particularly small and medium-sized businesses as they enter international markets or expand there. As a public-private organization and part of the Ministry of Employment and the Economy Group (with about 550 members), it helps its business customers see where there are opportunities for niche technologies and where new commercial and marketing skills would be needed to reposition Finnish products and technological ingenuity. As has already been referenced in the previous chapter, SITRA is an independent public foundation that operates under the supervision of the Finnish Parliament and is dedicated to promoting the economic prosperity and future success of Finland. Its current program is focused on issues of health care, food and nutrition, energy and network-based operations in the machine and metal industries. SITRA has continued to highlight the central importance of providing people with better health care services and empowering them to understand their health as well as take better care of themselves. Its experience in business development and venture capital investment also complements the work of other organizations such as Tekes.

For more than a century, the Helsinki University of Technology[14] (Finland's oldest and largest university of technology and architecture), has both boosted the country's technological development and led the way in building the basis for Finland's economic dynamism and high standard of living. Its Otaniemi International Innovation Center (OIIC) offers a range of expert services, from research liaison on R & D projects to assistance with patents and setting up companies. In promoting the development and use of competencies, skills and research findings, OIIC acts as a one-stop shop and a hub for the universities R & D services, fostering innovation activities both within the campus area and more widely throughout Great Helsinki. Other research institutes and scientific centers of excellence that are part of the OIIC science park include the Technological Geological Survey of Finland, Lifelong Learning Institute (Dipoli), Innopoli (a research and commercial development center) and Culminatum (the Helsinki Region Center of Excellence). The role of Culminatum[15] is to improve cooperation and collaboration among local municipalities, universities, and the business community in the Helsinki region and to help in developing resources and cutting-edge competencies that would enable the region to succeed in an ever-more competitive world. It focuses on creating the tools for high-tech businesses to develop and flourish, and to smoothing the path of research knowledge and know-how to its application in business and industry. For Culminatum, there is the constant reminder of a simple truism that nothing ever happens automatically; excellence needs to be nurtured and developed every day.

There is also the *BioTurku*—a conglomerate of companies,[16] research units and academic institutions that have joined forces to work together in the field of drug development, diagnostics, biomaterials and functional foods. The Turku region (Southwest Finland) is the country's major biomedical cluster with over half of the country's pharmaceutical and diagnostic industry—among them well-known multinationals. In the Turku region is also *DigiTurku*—a cluster of organizations focused on ICT research, expertise and business development,[17] employing nearly 15,000 people and an annual turnover of approximately €8.7 billion. Beyond the traditional emphasis of ICT on electronics and content applications in mobile communications, a developing trend has been the coming together of software and telecommunications experts, on the one hand, and biotech and information technology experts, on the other, to create new convergent knowledge and skills in *bioinformatics*.

Overall, Finland has built a strong set of technology clusters in its inventory: ICT, engineering, forest products and energy. "With world-leading expertise, know-how and a strong corporate sector in these areas, the challenge now is deciding which field Finland realistically wants to be the best in, in a world where excellence is certainly not in short supply."[18] This is particularly important because "retaining and further developing the dynamism of these clusters is critical to their future; and also critical to using them as foundation for pushing forward developments in related fields and more distant ones as well."[19] In the thinking that undergirds the Finnish innovation system, the future is now, and the more distant discoveries might already be around the corner.

There already exist innovations meant to develop more power from the wind, renewable diesel that has 40–60% less greenhouse emissions than conventional fossil fuels (Neste Oil's NexBTL bio-diesel produced from feedstock including waste animal fat), ethanol from industrial waste, Enfucell's softBattery technology, electric vehicle technologies, safe solutions for spent nuclear fuel (KBS-3 disposal concept) that can last for thousands of years, underground waste disposal, intelligent packing solutions, dual-action technology for healthier foods, thermally modified rot-free wood products, green solutions for an ecologically sustainable society (Cleantech Finland), smart cities (competitive urban infrastructure and user-friendly spaces), metallurgical technology, biofibre and plastic waste renewal, intelligent smoke alarms, eco-efficient construction, new advanced building elevators (KONE technology), cloud- computing and data analytics and more too numerous to mention.

4.2.1 Innovation Clusters and Networks as Learning Organizations

There is the ever-persistent tendency to limit our conceptualization of innovation as a process of change that leads to new ways of doing things, but one is also mindful of the fact that innovation also involves, more than anything else, a process of inquiry and discovery. This process can be facilitated either by making marginal adjustments in the exiting production structure or technology or by embarking on a more radical change in the idea, logic and philosophical platform that undergird the production structure. Either way and to a certain degree, innovation involves an element of *creative destruction* in which core ideas are retained as building blocks and foundation for the generation of new knowledge. In situations where core ideas are totally discarded and new knowledge developed from scratch, what we have may be partly innovation but also a form of discovery. Discovery seeks to distribute innovative capacity throughout a network or knowledge clusters in such a way that the system can collectively adapt as well as leverage new learning and opportunities.

Although the traditional view of innovation is that it follows a linear model and is generally activated by latent or inertial forces within the market or society, it is a more complex and multidimensional phenomenon. Innovation is an interactive learning process whereby feedback effects occur among market, science and technology-related activities both within and between the private and public sectors.[20] As a means of creating growth, increasing productivity and employment, the Finnish model includes the creation of regional centers or science parks and network systems that encourage shared experiences in innovation and knowledge creation between firms and universities. Research by J.S. Jauhiainen[21] and K. Ala-Rämi[22] examine how innovation-oriented regional science centers in Northern Finland (particularly *Multipolis*) enhance cooperation among high-tech firms, research and education institutes and regional development actors. While Jauhiainen found that cooperation had a positive effect on high-tech firms' innovation activity, Ala-Rämi's study found that the use of ICT solutions was very important for making contacts and enjoying continued cooperation. In both cases, innovation activities involve continuous collaboration and the linkage of several functions such as applied research, development of prototypes, industrial investment and production start-ups, and market and product adaptation.[23] Because most firms are more interested in the mutual sharing of knowledge and experience that derive from cooperation within and between network systems, they are also likely to be more successful innovators.

Regional Innovation Policy: Incremental Adjustments and Radical Departures

As an exercise in the acquisition of new knowledge, an innovation is a form of organizational learning. It often involves making incremental adjustments at the margins or radical departures from the existing norm—an unfreezing process that could set in motion novel approaches to problem solving and better ways of doing things. Nonetheless, genuine learning and the ability to develop innovative breakthrough products and services have to go beyond the collection and processing of information; they must also embrace the creation of insight and knowledge.[24] By this, organizations *unlearn* old habits as they *re-learn* new ways of doing things as well as new processes and their potential or actual effects on organizational outcomes. This type of learning is reminiscent of Argyris and Schon's[25] concept of *double-loop learning*—which generally results in a change in the *theory-in-use* (how organizations actually do things). In the case of double-loop learning, the values, strategies and assumptions that govern organizational action are changed to create a more efficient environment. It involves a drastic rethink of the existing operational norm so that adverse consequences can be avoided, or at best mitigated, in the future. In situations where double-loop learning is insufficient for dealing with the more deeper cultural and process issues or where the status quo seems more resilient to innovation and change, then *deuterolearning*—learning about improving the learning system itself (*learning to learn*), including the structural and behavioral components that determine how learning takes place, become more expedient. While value creation is at the core of the innovation process: paradoxes, trial and error, celebrating mistakes, embracing disagreements and conflict situations or floating trial balloons can, in and of themselves, create positive conditions for the exercise of new and constructive learning. In a similar vein and to the extent that it offers novel approaches in dealing with structural and process issues within the organization, the idea of *learning to learn* is also a form of innovation.

For reasons of inquiry and discovery, successful innovations seek to challenge old assumptions, encourage new practices and create a new attitude in the way organizational members internalize and go about their routine activities. This has been one of the key objectives of the Jyväskylä Regional Science Park (JRSP) in Northern Finland. It was created to develop and strengthen businesses based on new knowledge and technology but also to assist in building a favorable operational environment for high-tech companies in the regional area of Northern Finland. In addition to having responsibility for the Center of Expertise program, the JRSP is also instrumental in the growth of the ICT cluster that has remained one of the local economy's key features in the Jyväskylä urban region. The other regional science park is the *Multipolis*[26]—a publicly funded innovation-oriented high-technology network that connects the highly developed urban region of Oulu (technopolis) and the rest of the less-developed areas of northern Finland (learning regions). As a result of difference in geographic, demographic and economic conditions in northern Finland, the technology policy encompasses two models of regional innovation systems.[27] The first concept is the technopolis, which focuses on the development of high-technology and knowledge-intensive fields and enterprises in larger *functional urban regions* (FUR). The second is the Multipolis, a concept based on "learning regions" that cover broad fundamentals of development such as education, training and enhancing public sector institutions in innovation policies.[28]

By linking high-technology enterprises and specialized technology clusters, regional developers, higher education and research institutes, the *Multipolis* framework has been able to leverage the enhanced cooperation, exchange of new information, and social networking in the facilitation of common research projects and product development. The bases for the innovation-based regional policy are to increase knowledge and competences in addition to new technologies, education and related training activities across the board. In light of the existing Regional Development Law, it offers a means of strengthening the learning capacity of each region in the country, lessening regional differences as well as supporting balanced development between the core (urban) and peripheral (rural) areas.[29] Table 4.1 provides an illustrative account of the *Multipolis* network,[30] which connects high-technology enterprises in and among 15 specialized high-technology clusters (polis), regional developers and higher education and research institutes located all around in northern Finland.

While the activities cross several administrative, regional and functional borders, they also include universities and several regional development organizations in joint public-private projects or indirectly as providers of labor to the Multipolis network. On the one hand, Finland's national technology and innovation policy reflects a conscious and systematic effort to embed technology in a wider socioeconomic framework as a means of achieving equality of development; on the other, it addresses the unique challenges and opportunities faced by each locality or region. Figure 4.6 is indicative of the fact that some of the

Table 4.1 Multipolis Network Depicting Technology Clusters (Polis) and Specific Fields of Expertise

Name (Polis)/ Technology Clusters	*Municipality in Northern Finland*	*Founding Year*	*No. of Enterprises/ High-Tech Firms*	*No. of Workers Employed*	*Specific Field of Technology*
Astropolis	Sodankylä	2000	7/7	50	Electronics, precision measurement
Aurora Borealis	Rovaniemi	2001	40/40	1000	Winter and cold technology
Digipolis	Kemi	1989	30/15	400	Electronics, data processing
Envitecpolis	Iisalmi	2003	33/16	330	Environmental technology
Humanpolis	Utajärvi	2000	5/1	20	Welfare, environmental technology
Measurepolis	Kajaani	1998	12/9	800	Precision measurement, electronics
Mediapolis	Tornio	2002	20/15	200	Content production, media
Micropolis	Ii	2002	18/4	100	Micro and nano-technology
Naturpolis	Kuusamo	1993	15/8	90	Program industries
Netpolis	Kaustinen-Kokkola	2001	20/10	300	Program industries
Powerpolis	Kemijärvi	2001	3/1	40	Electronics, data processing
RFM-polis	Oulun eteläinen	1998	40/40	2000	Data, media, welfare technology
Softpolis	Raahe	1993	19/14	ND	Program industries, data
Steelpolis	Raahe	2001	30/3	4000	Metal technology
Technopolis	Oulu	1982	190/220	6000	Micro, nano, welfare, bio, environmental technologies

Source: Adapted from Jussi S. Jauhiainen. 2006. Multipolis: High-technology Network in Northern Finland, p. 1416.

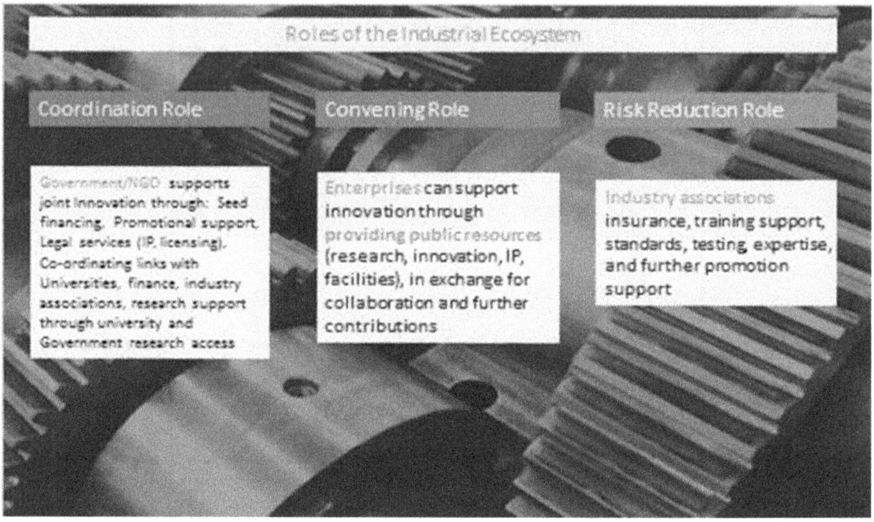

Figure 4.6 An Innovation Ecosystem for Emerging Industries.

organizational processes and functional roles that sustain the regional science parks and technology centers are similar to those of an industrial ecosystem.

There are three main goals in an innovation eco-system: *coordination role,* played by government and non-governmental organizations; *convening role,* played by enterprises through business linkages, IP, and value chains and *risk reduction role,* usually played by government or industrial associations, private support and promotional incentives. Finnish innovation success is a result of a public-private partnership where the government plays a coordinating and financing role, while enterprises provide the skill-sets needed for various research outputs and associational groups and labor serve in a supporting and facilitating role. While universities have played a central role in the innovation process, they should also be seen as critical especially for local economic growth and development. In addition to exploiting their laboratory discoveries by patenting and licensing intellectual property to local firms, they can also help to attract new knowledge and financial resources as well as adapt knowledge originating elsewhere to local conditions.[31] Universities should remain alert to the kinds of industrial transformation (new industry formations, industry transplantation, industry diversification or industry upgrading) occurring in their local communities, since these often demand different patterns of technology take-ups that require unique sets of skills and knowledge.[32] Hence, a strategic approach to local economic development is the development of a stronger awareness of the pathways along which local industries are evolving, as well as the innovation processes associated with those pathways,[33] so that resources and existing know-how can be adapted in a timely manner to take advantage of the potential opportunities.

4.2.2 Innovation Quality

It is common knowledge that on a performance level, Finland ranks very high on the Global Innovation Index (GII), which consists of a ranking of world economies' innovation capabilities and results. Published annually since 2007, the GII has become a chief benchmarking tool for business executives, policy makers and others seeking insight into the state of innovation around the world. To boost a country's innovation capabilities[34] oftentimes requires the creation of conditions that allow enterprise champions, state-owned enterprises, family-owned conglomerates, government, multinational corporations, industry associations and labor to become the critical drivers of innovation hub (technopolis) activities (see Figure 4.7). These enterprise champions (industry or product leaders) are able to facilitate the building of talent pools that stimulate innovation as well as help to bridge the gap between research and commercial success.

The GII is calculated as the average of two sub-indices:[35] the input sub-index and the output sub-index. The input sub-index measures elements of the national economy, which embody innovative activities grouped into five pillars: institutions, human capital and research, infrastructure, market sophistication and business sophistication. On the other hand, the output sub-index captures actual

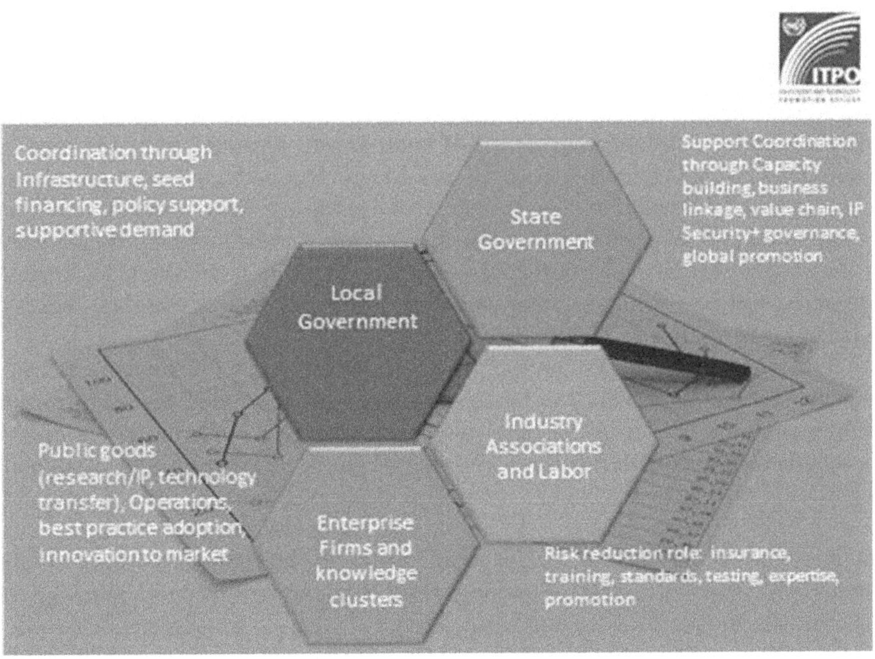

Figure 4.7 Public-Private Partnership: State Dirigisme and Multi-Platform Project Model.

evidence of innovation results measured in two pillars: knowledge and technology outputs and creative outputs. Based on a country's ranking on the Global Innovation Index, we can calculate the innovation efficiency as a ratio of output sub-index over input sub-index, viz:

> *Innovation Efficiency (Ratio)* = *Output Sub-Index divided by Input Sub-Index*
> Where *Input Sub-Index* = Institutions, Human capital, Infrastructure, Market sophistication, and Business sophistication; while
> *Output Sub-Index* = Knowledge and technology outputs, and Creative outputs.

On the Global Innovation Index, Finland ranked 4th in 2012, 6th in 2013, and 4th in 2014 with a relative score of 60.7 and percentage rank of 0.98. The specific criteria for measuring the input and output sub-indexes can be readily accessed from the Global Innovation Report 2014 or online at https://www.globalinnovationindex.org.

4.3 Technology and the Future

There is little doubt that technology has brought about dramatic changes to human society and life itself; it has also brought about efficiency in the work process and improvements in health and human welfare. The natural state of mankind, to the extent that it ever existed, has changed into an artificially constructed organic system such that the possibility of reversing to its original and pristine nature has been lost for good. Human society has become a captive audience. History reveals that technological futurism is largely unchanging over time, while the present vision of the future continues to display a startling and unselfconscious world of automatism, linearity, and precision.[36] Amid increasing tension between the need to control and our wish to preserve our values, the heightened prominence of technology in society makes the interrelated task of profiting from its opportunities and containing its dangers a major intellectual and political challenge.[37] What is technology's carrying capacity, and how far can it take us before it reaches critical mass and begins to retrogress or atrophy?

While it remains rather unclear where technologies of the future will take human society, if the past can be seen as prologue, the future seems pregnant with many possibilities. How these possibilities actually shake out in the future is relatively unknown at this time. Through its heavy investment in research and knowledge management, Finland remains at the cusp of emerging discoveries and possibilities in the technologies of the future. The moral behind this initiative is premised on an uncanny belief that pushing back the technological frontier is all about innovation, new ways of looking at old problems or new ones, believing in an idea and finding a way to put it into practice.[38] There is a complementary relationship between a strong technological foundation and economic growth; both are critical in laying a solid foundation for future prosperity and social well-being. Finnish technology policy has continued to push beyond the boundaries in the development of new solutions in areas as diverse as "digital content, e-commerce, e-learning, financial management, Internet connectivity, neural networks, smart cards, spatial information systems, and in nurturing new mobile applications and services."[39]

The perennial focus on a healthier future, aging and quality of life issues has continued to engender advanced and cutting-edge research and development in the field of biosciences and biotechnology—from genetic profiling, biochemical analysis and drug targeting, research into vaccines and immunotherapeutics. "Advances made in areas such as enzymes and drug discovery are opening up new possibilities for more efficient industrial processes, diagnostic systems, and safer and more efficient therapies for various chronic diseases."[40] The ongoing national campaign on lifestyle changes, physical fitness, food and nutrition are also part of a broader policy regime geared toward advancing quality of life as well as preventive medicine. But this effort has not stopped in the areas of health care or bioscience; it essentially covers all other areas that can contribute in fundamental ways to improvements in human and social well-being.

The forestry and wood and pulp products industry comes to mind. The very size of Finland's forest products industry and its relevance to the national economy has been reflected in a steady stream of new systems, not only for improving the mechanical manufacturing aspects of pulp and papermaking and wood processing, but also the chemistry of the processes involved. The resultant effect is that new composite materials, consumer and industrial products, biofuel and chemical resins hardly imagined a few years ago now have the possibility of being produced from wood and pulp products. Table 4.2 provides a list of products and possibilities that can be derived from wood and pulp products. They include wood fiber materials, composite materials, manufactured wood products and chemically processed goods.

By culture and character, the Finns are a very inventive people.[41] It is well understood that creativity, inventiveness and innovativeness are essential in every sector of the economy, in helping underpin competiveness, entrepreneurship and export growth. For example, about one-third of Finnish inventions are generated by private individuals and university researchers, one-third by small companies and one-third by larger companies. "A recent survey of 600 innovations showed that 66% were totally new ideas and 31% represented significant improvements on existing products."[42] On a per capita basis, this is a monumental achievement for a country the size of Finland. Furthermore, continuous innovation provides a means for diversifying the economy and the production structure that undergirds it, overcoming the competition plus creating new avenues for employment and tax revenues accruable to the state. In the long run, countries with as much *innovation capital* as Finland are more likely to be in a stronger position to address contingent issues of sustainable development, investment and maintenance of a high standard of living. Although its technology policy has been premised on the building of linkages among different *knowledge institutions* in the innovation

Table 4.2 Products of the Future Derived from Wood and Wood Pulp

Wood Fiber Materials *[Kuitumateriaaleja]*	*Manufactured Wood Products* *[Puujalosteita]*	*Composite Materials* *[Komposiiteja]*	*Chemically Processed Goods* *[Kemiallisia jalosteita]*
Furnishing fabrics made out of paper yarn	Exterior cladding for thermo-wood	Casts made of composite wood	Electricity produced from wood chips
Viscose made out of cellulose	Supporting structures of cross-laminated timber	Furniture and loudspeakers made of composite wood	Wood care ointment produced from spruce resin
Electronic paper made out of wood	Wooden frames for eye glasses	Musical instruments made of natural fiber composites	Cancer drug made from spruce shoots
Diapers made out of cellulose	Eco-friendly wooden furniture		Biofuel made from waste wood
Biodegradable cups for hot and cold drinks	Lamp shades made of molded birchwood		
Biodegradable wrapping for foodstuffs	Outdoor elements (porches, quays, etc.) made of modified wood		

Source: Compiled by author based on information from publication sources of the Parliament of Finland.

eco-system (such as Aalto University and others) as a means of facilitating the propagation or percolation of inventions,[43] a more nuanced objective is that of technological diffusion on the one hand and market domination, on the other.

4.3.1 Of Futurisms and New Directions

Today, we witness the resulting convergence of three elements: politics, economy and technology. Everyone is learning from everyone everywhere because new and advanced technologies have broken down the walls of distance. The building of the computer in 1946 and the consequent birth of the Internet and the World Wide Web years latter have led to new possibilities, new horizons and new ways of life. No one or community has been spared. The world of organizations and businesses continues to witness a dramatic shift from traditional hierarchical arrangements to lateral or horizontal formations. We are now at the end of automation but witnessing the beginning of Industrialization 4.0 (the new Industrial Revolution)—the idea of intelligent machines communicating and interacting alongside with intelligent production. By integrating new technologies into products and processes, the new industrial revolution takes automation and smart connectivity to a much higher level. It is now the new normal in which intelligent robots in smart factories have become the major drivers of the digital economy.

As revolutionary as it seems, Industrialization 4.0 will demand new processes and skill-sets in dealing with robots manufacturing other robots as well as the complexity of goods in charge of their own production. Mobility patterns will be enhanced by the ability to reconfigure communication and technical linkages. What has generally been known is that "as innovation widens across populations, it tends to prompt the emergence of new sub-populations and niches which then confront each other in new eco-systems of competition and domination."[44] Hence, the evolving *knowledge economy* will necessitate increasing volatility and restructuring of social and economic relations both in terms of distribution and consumption patterns and in the emergence of novel organizational architectures. Colonization of futuristic technological fields, whether at the micro (market-led) or macro (state-led) level, will depend on the distinctive institutional, technological and political path dependencies involved in its own socio-economic development considered in light of their abrasion against those of the international system.[45]

Hence, as creative innovations become diffused and taken up across the wider society, they are more likely to abrade against larger crosscurrents of social and economic change, further tipping them in new and uncertain directions.[46] In response to global challenges, we stand to be inundated with exotic and even romantic technological solutions ranging from smart health care, smart city solutions, smart old-age support and advanced data analytics, to predictive decision support systems. Technology is continuously being pushed onto almost everything about life and the human society, but whether all these make our finite world better or more livable is a central debate among futures scholars. As Timo Turja, Chief Information specialist at the Library of the Finnish Parliament, pointed out in his speech to the 28th IFLA *Pre-Conference of Library and Research Services for Parliaments*, August 8–10, 2012, Helsinki: "We members of Parliament respect science, but we must remember that we cannot follow the opinions of the scientist without question. If scientists had had their way, no one would today be travelling by train."

Notes

1 Wiki Countries Navigation. 2015.
2 Ibid.
3 Tahvannainen, A-J. 2011. Diagnosing the growing pains of a technology-based industry: An examination of the Finnish biotechnology industry in light of empirical economics. Helsinki, Finland: *The Research Institute of the Finnish Economy*, 18–40, p. 18.
4 Ibid.
5 Finnish Communications Regulatory Authority (FICORA). 2011. The revolution in communications: Will all consumers make the switch to the Internet? Helsinki, Finland: *FICORA*, p. 5.
6 Finnish Academies of Technology. 2004. High technology Finland 2004. Helsinki, Finland: Finnish Academies of Technology, p. 23.

7 Ibid.
8 Herring, P. 2013. HighTech Finland 2013. Helsinki, Finland: HighTech Finland, p. 8.
9 HighTech Finland. 2008. A world of ideas and innovation. Helsinki, Finland: HighTech Finland, p. 10.
10 Statement credited to E. Virtanen, Permanent Secretary at the Ministry of Employment and Economy. In HighTech Finland 2008, p. 12.
11 Herring, P. 2004. High technology Finland 2004. Helsinki, Finland: Finnish Academies of Technology, p. 10.
12 Ibid.
13 HighTech Finland. A world of ideas and innovation, pp. 9–15.
14 Ilmavirta, V. 2004. High technology Finland 2004: An essential link in Finland's innovation chain. Helsinki, Finland: Finnish Academies of Technology, pp. 22–23.
15 Holstila, E. 2004. High technology Finland: Creative people are the critical success factor. Helsinki, Finland: Finnish Academies of Technology, p. 29.
16 Sorakunas, E. 2004. HighTechnology Finland. 2004: Rooted in history, looking to the future. Helsinki, Finland: Finnish Academies of Technology, p. 26.
17 Sorakunas, E. 2004. HighTechnology Finland. 2004: Rooted in history, looking to the future, p. 26.
18 HighTech Finland. A world of ideas and innovation. p. 11.
19 Ibid.
20 Mukkala, K. 2010. The role of regional policies in promoting networking and innovative activity: Evidence from small Finnish high-tech firms. *European Planning Studies*, Vol, 18(7): 1059.
21 Jauhiainen, J. S. 2006. Multipolis: High technology networks in Northern Finland. *European Planning Studies*, 14(10): 1407–1428.
22 Ala-Rämi, K. 2007. Communication and distance in collaboration between high technology enterprises in Northern Finland, *European Planning Studies*, 15(8): 1047–1062.
23 Patrucco, P. P. 2003. Institutional variety, networking and knowledge exchange: Communication and innovation in the case of Brianza technological district. *Regional Studies*, 37(2): 159–172. Also see Jauhiainen, J. S. Multipolis: High technology networks in Northern Finland, pp. 1407–1428; and Ala-Rämi, K. Communication and distance in collaboration between high technology enterprises in Northern Finland, pp. 1047–1062.
24 Morgan, G. 2006. *Images of organization*. Thousand Oaks, CA: Sage Publications, p. 88.
25 Argyris, C. and D. Schon. 1996. Organizational learning II: *Theory, method and Practice*. Reading, MA: Addison Wesley.
26 Jauhiainen, J. S. Multipolis: High-technology network in Northern Finland, pp. 1407–1428, p. 1407.
27 Jauhiainen, J. S. Multipolis: High-technology network in Northern Finland, p. 1417.
28 Kautonen, M. and J. Kolehmainen. 2001. Perspectives on regional innovation policy of the learning economy. In *Alueiden kilpailukyvyn kahdeksan elementtiä*, ed. M. Sotarauta and N. Mustikkamäki, 109–128. Helsinki, Finland: Suomen Kuntaliitto.
29 Jauhiainen, J. S. Multipolis: High-technology network in Northern Finland, p. 1408.
30 Ibid. p. 1415.
31 Lester, R. K. and M. Sotarauta. 2007. Executive summary: Innovation, universities, and the competitiveness of regions, *Technology review 214*. Helsinki, Finland: Tekes, p. 2.
32 Ibid.
33 Ibid. p. 3.
34 Statement credited to Barry Jaruzelski, Senior Partner at Booz & Company. http://www.wipo.int/pressroom/en/articles/2013/article_0016.html. Accessed on August 10, 2015.

35 World Intellectual Property Organization (WIPO). 2013. Global innovation index 2013. http://www.wipo.int/pressroom/en/articles/2013/article_0016.html, 6.
36 Edgerton, D. 2009. The shock of the old. In *Technology and the future*, ed. A. H. Teich, 63–73. Boston, MA: Wadsworth, p. 67.
37 Mesthene, E. G. 2009. The role of technology in society. In *Technology and the future*, ed. A. H. Teich, 77–86. Boston, MA: Wadsworth, pp. 80–81.
38 Herring, P. 2003. Inspired by ideas, p. 11.
39 High Technology Finland. Communications and information, p. 32.
40 Herring, P. 2003. Inspired by ideas, pp. 11–12.
41 Sipilä, K. 2003. Inventions create a better future. *High technology Finland* 2003. Helsinki, Finland: High Technology Finland, p. 15.
42 Ibid.
43 Room, G. 2011. *Complexity, institutions, and public policy: Agile decision-making in a turbulent world.* Cheltenham, UK: Edward Elgar, p. 161.
44 Ibid. p. 279.
45 Ibid. p. 283.
46 Ibid. p. 308.

5 Finnish Lessons

Nurturing Education

Finland ranks at the top of most major indicators of education and literacy in the world. With a literacy rate of 100% and no gender differences or male/female distinctions, it has managed in a relatively short time to achieve what many other countries have been unable to do for decades. It is considered one of the most educated and wealthy countries in the world with a highly skilled labor force, a robust health care system and a high standard of living. It has also managed to lay an enviable foundation of sustained social and economic development, partly premised on advancements and an emphasis on cutting-edge information technology culture, a learning and educational system that provides the creative energy and innovation for future growth and development.

Basic studies in the humanities as well as the sciences are equally appreciated as part of the educational system, in such a way that 'hard' and 'soft' values are in harmony and diverse cultural activities, supported from public funds, counterbalance science as an important part of a stable constitutional welfare state. "International studies now rate Finland as one of the world's most competitive and innovative countries, and foreign visitors come to learn of its ethical corporate culture."[1]

Since the country implemented huge education reforms more than 40 years ago, the school system has consistently been at the top of all international rankings for education systems,[2] and it has the numbers to show for it. The often-asked question about the reasons for Finland's educational achievement seems to mask two central elements at its core: the political will of its national leaders and a willingness to invest in its people. In this regard, education policy essentially follows the broad premises set by government with a primary aim of reducing inequalities in society—an objective that also draws from its secular Lutheran underpinnings. This is a tradition that the country has embraced for years even before independence from the Soviet Union in 1917; hence, what we see today are the cumulative effects of that investment and its spillover into several facets of the national economy and the social system. In addition, basic cultural developments in the country have also spurred innovations in the education sector. Among these is a culture of caring, leadership skills, a well-educated teaching staff and creativity in harnessing natural resources leading to increased investment in engineering education.

5.1 Bringing the State Back In: Policy as National Will

The objectives of Finnish education policy are based on quality, efficiency, equality for all and internationalization. It is seen as playing a supportive role to the competitiveness of the country's welfare state while assuring its cultural, social and economic sustainability. The Constitution grants education as a basic right;[3] hence, it behooves public authorities to secure equal opportunities for every resident of Finland to get an education. Legislation provides for compulsory schooling and the right to free pre-primary and basic education, including post-graduate education in universities. In this regard, it seeks to provide equal opportunities for learning and growth to every pupil or student, as well as provide the needed supporting mechanisms such as mitigating physical, attitudinal or pedagogical barriers to learning by means of early intervention. Basic education is free and by virtue of the *Study Allowance Act* of 1972, students enrolled in secondary schools and at universities are entitled to financial aid in the form of allowances and/or low-interest loans.

As illustrated in the OECD's Education Policy Outlook,[4] governance of the Finnish education system is decentralized with a significant level of authority, responsibility and decision making granted to the local governments, particularly to municipalities or joint municipal authorities. Local authorities have responsibility for organizing basic education, deciding on funding allocations, designing and implementing local curriculum, granting scope of autonomy to local schools and recruiting and training personnel. At the higher education level, the *Universities Act* of 2009 grants further administrative and financial autonomy to universities. Finnish universities (including Polytechnics) and the Ministry of Education and Culture have established performance agreements that define operational and qualitative performance targets for each university or polytechnic, and these provide the general basis for the university's student intake, selection, curricula and instruction.

In fundamental ways, the key to success has been the solid preparation given to students at early childhood and in basic education and also the degree of independence and autonomy granted to teachers in curriculum development as well as pedagogic style. The School Systems Act, which was passed in 1968, abolished the two-track elementary school system and replaced it with a comprehensive system that would take nine years to complete. It also created a uniform educational system throughout the country and brought the existing private schools into the comprehensive system. It sought to create a flexible school system that not only matched individual qualities but also their aspirations and talents. As further incentive,[5] each child who completes primary school is granted a place of study in general upper education, vocational or apprenticeship levels, while applicants who have completed basic education or have no secondary level degree or place of study are given priority in secondary education student selection. About 75% of education funding in Finland is provided through social security. The state steps in to help poorer and less able local governments to make up for the funding needed to execute their national education objectives.

5.1.1 Core Principles and Principal Actors

The end of WW II saw education take up center stage in the national quest for political, economic and social transformation. While private schools had operated independently, a significant social innovation in 1950 was the passing of legislation that provided them with state subsidies, but invariably extended government control over them.[6] The intent was to achieve equal educational opportunity as well as a unified curriculum across Finland, which as envisaged, would allow for equity and consistency in outcomes. This will assure that equal educational access is available to all irrespective of socioeconomic status, ability to pay or other ascriptive qualities such as ethnicity or gender. There is a universal free education that also encourages lifelong learning even after formal education has ended. Formal education is seen as part of a much larger learning process that continues beyond one's earlier productive years.[7]

Although the objectives of the Government Programme are built around the Europe 2020 Strategy,[8] state policy on education and research is set by Parliament in conjunction with the government, and these are articulated in the *Education and Research Development Plan,* which is adopted every four years as the basis for implementation. The document covers all forms of education from early childhood to adult education as well as research conducted in universities and polytechnics. While various agencies play a role in the implementation of education and research policy, there are some whose functions stand out from the rest. Figure 5.1 provides the administrative architecture that supports

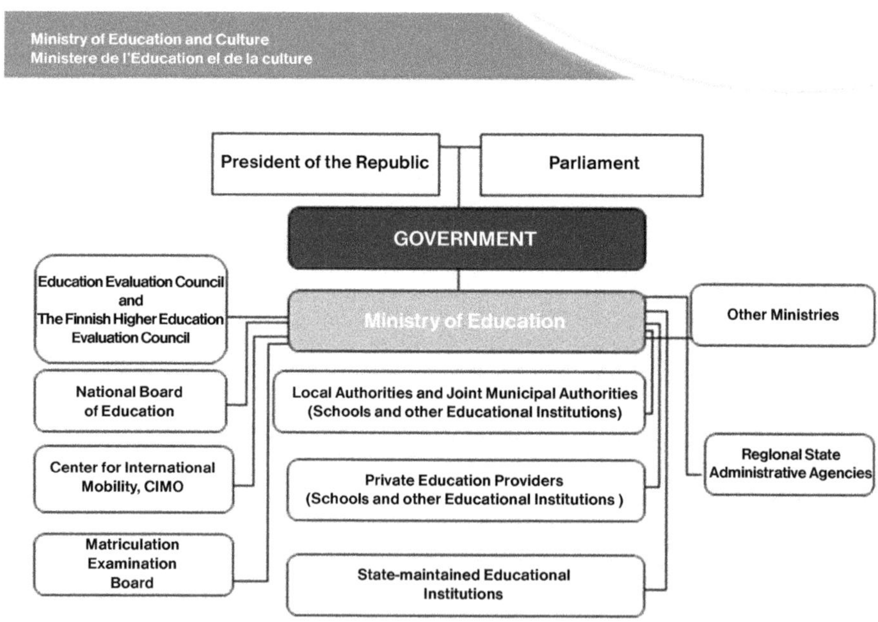

Figure 5.1 Administration of Basic and Upper Secondary Education in Finland.

Finnish education and research in terms of policy making and implementation. The Ministry of Education and Culture[9] is responsible for developing educational, scientific, cultural, sports and youth policies, while the Finnish National Board of Education, though under the Ministry of Education and Culture, works with the implementation, monitoring and overseeing developments and innovations in the educational sector, particularly the national core curricula for pre-primary, basic and general upper secondary education as well as adult education and training. It is also responsible for deciding on educational or employment eligibility conferred by foreign qualifications for a post or for further studies in Finland.

There is also the Finnish Education Evaluation Center (FINEEC)—an independent government agency responsible for the national evaluation of educational performance from early childhood to higher education. The Finnish ENIC/NARIC (*European Network of National Information Centers on Academic Recognition and Mobility/Network of National Academic Recognition Information Centers*), the Academy of Finland as major funding agency, and the Center for International Mobility (CIMO)—a sub-agency of the Ministry of Education and Culture that offers services and expertise to encourage cross-cultural communication as well as administer the Study in Finland program. CIMO is also the national agency for the EU's education, training and youth programs and also plays a central role in international mobility and cooperative programs that support Finnish and international studies.

There are other stakeholder groups that play central roles in education policy such as the Association of Finnish Local and Regional Authorities, the Confederation of Finnish Industries, the Central Organization of Finnish Trade Unions, the Finnish Confederation of Professionals, the Confederation of Unions for Professional and Managerial Staff in Finland, the Trade Union of Education in Finland, student unions and education providers. The multiplicity of stakeholders in the decision-making process reinforces the democratic principles of inclusiveness, broad participation, consultation and the rather uncommon virtue that good ideas are not an exclusive preserve but can come from anywhere or from anybody.

While evidence points to the Ministry of Education and Culture as playing a key role in the administration of education policy, the functional design of its four major departments and the specialized units within them makes the role of the Permanent Secretary critical (Figure 5.2). By facilitating horizontal communication between the departments, the existing division of labor enables continuous development of expertise and shared knowledge. One of its primary focuses is the continuous improvement and evaluation in higher education.

Unlike the American educational system, the Finnish model achieves three important values, that of equity, access and excellence. While the state provides most of the financing, support, and control consistent with the Nordic model of the welfare state, the American system has been one in "which the private

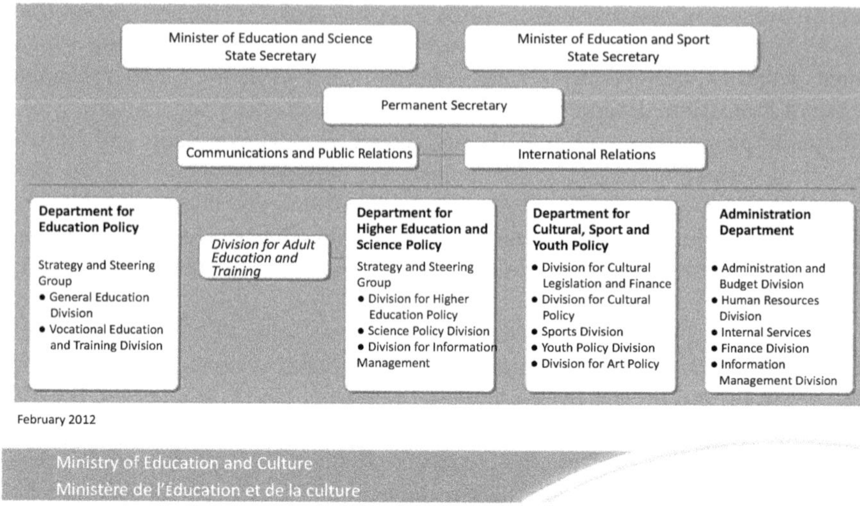

Figure 5.2 Ministry of Education and Culture Showing Specialized Departments and Divisions.

sector provides a large part of the education and individuals pay for most of their tuition." But how much value do both models get for the amount of money and resources expended? As pointed out by *The Economist*, "tests suggest that many students in the United States do not learn enough, and they seem to work less on their academic duties than they used to."[10] One would therefore expect the outcome to be less rosy or less encouraging compared to the investment in time, money and effort. The conclusion is that the average performance of "American graduates compared with those of other countries has continued to decline; while at the social level, it shows that higher education on average has not increased social mobility but has been more likely to reinforce existing barriers."[11]

In this context, Finland's experience with its graduates and their academic performance is the reverse of America's. The national political will and the state's proactive role in instituting a flexible policy that sees education as a right rather than a privilege meant that the values of equity, access and excellence can be more effectively achieved through state mechanisms as opposed to subjecting them to the exigencies of the market or the private sector. Although most of the schools are owned and organized by municipalities, they are funded by the state government based on a formula, and in most cases parents choose which school their kids attend. All costs are itemized each year and sent to the Ministry of Education and Culture, then divided evenly between the state (40%) and the municipalities (40%). The Ministry receives an annual budget for basic education, but distributes it on a monthly basis to the municipalities since they are the ones that actually conduct the business of education. However, the economic condition and the per capita income of each of the municipalities are taken into consideration, in such a way that small and less-disposed municipalities in rural and relatively poor areas can get additional funding.

5.2 Functionalism and the Structure of Pedagogy

The profile of the Finnish education system follows a truly functionalist model, in the sense that several distinct roles are linked one to the other in such a way that they contribute to a functional whole. It also demonstrates how a series of activities, practices and differentiated roles are able to sustain a level of organic and functional equilibrium over time.[12] While equilibrium suggests a type of 'stable state,' it also points to the fact that such a 'state' can only be sustained if the various system parts perform their functions as designed. Consistent with Talcott Parson's structural-functional framework,[13] one can also see the education system as a cultural embodiment of the national will, whereby individual citizens have come to internalize shared norms and understanding in such a way that they become the central basis for national policy legitimation.

This is an important factor in the public buy-in and almost universal acceptance of core principles such as equality, equity, access, uniformity and free tuition that undergird education in Finland. Because the argument and its justification resonate quite well in the body politic, its implementation thus becomes everybody's responsibility. Not only does the normative argument serve to legitimize the established functional patterns of operation or structure, which are necessary to implement the education policy, it has also become a model that the people want preserved or exported to other countries in the world. While the structure of the education system reflects these values, it offers a very flexible system of instruction in such a way that there are no 'dead-ends' that could prevent an individual from advancing from one level to another.

The Finnish Board of Education offers, in a nutshell, the essential features that have sustained the education system over the years, thereby making it consistently functional. Because the primary focus is on learning rather than testing, there are no national tests for pupils in basic education; instead, teachers are responsible for the assessments done in their respective subject areas following the objectives and stipulations in the curriculum. There is only the national matriculation examination held at the end of the general and upper secondary education (19th year), which (in addition to results from entrance tests) form the basis for admission to higher education (universities and polytechnics). There are no requirements for the Scholastic Aptitude Test (SAT) or the equivalent for the American College Test (ACT). All that is needed is the results from the matriculation examination, the school leaving certificate, and the results from the entrance test administered by the university one wishes to attend. Selection or admission is based on a combination of the score from the matriculation examination and the university test. Currently, there is a plan to digitize the matriculation examination at the upper secondary school level, and this would begin in 2016 and may take up to 2019 for its completion.

5.2.1 Enabling Structure

The Finnish educational system is grounded in four major categories that build upon each other as an evolutionary pathway (Figure 5.3). It consists of:

1. One year of voluntary pre-primary education
2. Nine years of primary education (Basic/Comprehensive school)
3. Upper Secondary Education, which consists of general and vocational Education
4. Higher Education, in universities and polytechnics (universities of applied sciences) up to the doctorate or licenciate degrees

The pre-primary education, basic education (*peruskoulu*) and the upper secondary and vocational training (Levels 1–3, which takes approximately 19 years) complemented by early childhood education and before-and-after-school activities, form a coherent learning pathway that supports children's growth, development and well-being. Higher education is the preserve of universities and polytechnics, which offer scientific research and instruction, as well as practical

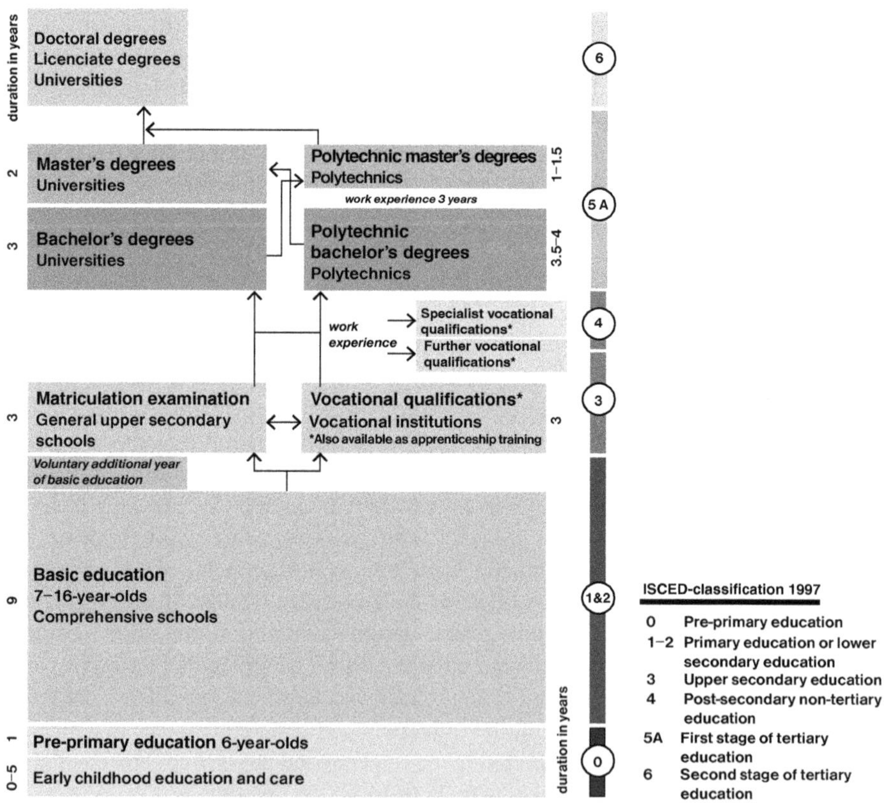

Figure 5.3 The Structure of Finland's Education System.

instructions in the applied sciences, respectively. In line with the principle of lifelong learning, adult education is provided at all education levels, either for the general education certificate, vocational qualifications or for taking courses that facilitate further skills in citizenship, recreational studies or other aspects of social and personal well-being.

5.2.2 Catching them Early: Focus on the Child

Recent statistics[14] from the OECD point to the fact that while the United States has fared quite well on provisions for the education of newborns to three-year-olds, when it comes to providing pre-school for three- to five-year-olds, it falls to the back of the class—behind Chile and just barely ahead of Lithuania and Greece. But in Finland, there is a different appeal and national commitment to early childhood and pre-primary education. While day care and pre-primary education are available and more than half of all four-year-olds participate, from 2001, pre-primary education was made available for six-year-olds with an emphasis on play and socializing as well as to prepare them for first grade of basic education. Today, Finnish law makes pre-primary education mandatory, and about 98% of young children are already involved in it. While it is available free of charge, transportation and day care be provided based on an income-assessed fee. "In addition, the state subsidizes parents, paying them approximately €150 euros per month for every child until he or she turns 17 years: and the schools provide food, medical care, counselling, and transportation where necessary."[15] Because it is a means of preparing children for school and future career prospects, the national core curriculum for pre-primary education, basic education, upper general and secondary education in the arts as well as curricula for preparatory education for immigrants is determined by the Finnish Board of Education. Pre-primary education is the responsibility of the respective municipalities, and even though parents do have a choice regarding where their kids go; all children are taught under the same educational system.

Basic Education (Comprehensive School)

In their 7th year, they enter into the basic education (Comprehensive School). The objective is to provide critical support for the children's growth and development and to endow them with the knowledge and skills needed in life. The basic school takes up to nine years to complete. During this time, they must take lessons in up to 18 compulsory subject matters that include languages, sciences (biology, chemistry, and physics), social sciences, music, sports and religion. Under basic education, there is free tuition, including textbooks and meals, and there is no tracking or streaming of students. The proportion of 15-year-olds[16] who have repeated at least a grade in 2009 was 2.8% and is among the lowest in OECD countries that have on average 13% of their students who repeat grade levels. Unlike the OECD, most of the variance in the performance of 15-year-olds is seen within schools rather than between schools, and this suggests that schools in Finland have similar achievement levels and that students are not selected or chosen based on academic ability. It also points to the fact that early intervention and individual guidance and support by all educational personnel have been crucial in ensuring that no one is left behind.

Upper Secondary Education and Vocational Education and Training (Vet)[17]

Upon graduation from upper secondary school or from a vocational training program, students are ready to enter into higher education or other tertiary institutions or they can opt to go directly to the job market if they have had vocational training. By a 2013 statistic, Finland had a high upper secondary education attainment rate of 90% (25 to 34 year olds) compared to the OECD average of 82%. At the end of the three-year general education curriculum (that is three additional years of study beyond basic school education), students can take the national matriculation examination, which provides access to higher or tertiary education. Because students can pursue vocational education training (VET) in upper secondary and polytechnics at the tertiary level of education, it can upon completion provide formal eligibility or qualification for entry into the labor market. Initial vocational training programs take three years to complete, including at least half a year of on-the-job training in various workplaces such as vocational institutions or apprenticeships. "Instead of a national matriculation examination, vocational students take school-level assessment of learning outcomes and a demonstration of skills to prove that they have achieved the vocational proficiency set out in the curriculum."[18]

As a way of reducing drop-out rates as well as making the system more flexible, a *modular structure* allows students to combine general education and vocational education and training studies at the same time. An OECD statistic[19] point to the fact that graduation rates in Finland's pre-vocational/vocational programs have been high in the neighborhood of 99% compared to the OECD average of 47% in 2011. While 66% of students go on to higher education (the highest rate in the European Union), the county spends about 30% less per student than the United States.[20] Nonetheless, administrators continue to address the need of ensuring that general education and vocational training programs are properly aligned to labor market needs and support effective transitions from program to work environment. *Adult education* is also supported as a way of extending individual working life, increase employment rate, improving productivity, increasing opportunities for leisure and encouraging life-long learning.

Higher Education and Tertiary Institutions

As indicated earlier, Finnish education policy allows individuals to attain the highest possible training or level they are able. Higher education is made up of universities and polytechnics; while the former emphasize scientific research and instructions, the later focuses on applied research. OECD statistic[21] indicates that over 39% of the population aged 25 to 34 years have attained higher or tertiary education. Although resource allocation and intake of students by field of study for tertiary education is based on forecasts for labor market demands, the expansion of tertiary education, nonetheless, implies providing a sufficiently broad array of studies to address the needs of the labor market as well as students' career interests. As in all other levels of the education system, there is no cost for higher education, and students can also get various financial aid packages. It should also be noted that the state's investment in higher education is also consistent with its strategy of facilitating networks of research and knowledge clusters in which universities and public-private research institutes, and industry, collaborate in the development of product and scientific innovations that contribute to economic growth and employment.

5.3 Teaching as Profession and Mentorship

In Finland,[22] teachers receive a strong theoretical and practical teacher education, are paid relatively well and are highly respected. Pre-service teacher training requirements in higher education institutions include a *competitive examination* to enter the training, which also includes a *teaching practicum*. All teachers in basic and general upper secondary education have a master's degree. As professionals, teachers have considerable pedagogical autonomy to interpret the curricula and to choose teaching methods and materials, as well as resources and methods for continuous assessment of students' progress. Continuing professional development (CPD), obligatory for most teachers, varies from one sector to the other and is partly defined in professional and trade union agreement. While employers (municipalities) decide on programs and the form of continuing professional development, local educational authorities are also required to fund at least three days of mandatory professional development per year for all teachers.

There is a broader recognition that funding for teacher's professional development is a worthy investment that would strengthen their overall capacity to deliver quality education to their students. While the teachers have to teach, they also need to develop the necessary skills to guide students on how to learn. They need to use very flexible methods so that every teaching and learning is student-centered. The need for flexibility has also encouraged the need for in-service training for teachers in the use of ICT to facilitate the learning process. Because his or her role has changed to that of helping students to develop the capacity to learn, the teacher now operates more like a moderator of the learning process: recasting learning environments to facilitate the needs of challenged pupils and changing and adjusting the mechanics of instructional delivery to advance the educational project and the process of *learning to learn*.

5.3.1 Evaluation and Curricula Development: Comparative Assessment and Quality Indicators

There are several ways to assess the extent to which the Finnish model has been successful in advancing the primary objective of its education policy. This includes students' classroom academic performance, ranking on national and international performance measurement indexes such as the Programme for International Student Assessment (PISA), subject matter competitive tests in the sciences, arts, and mathematics and overall tests in literacy to name but a few. The PISA provides a snapshot of the competence of students (in this case 15-year-olds) in reading, mathematics and sciences in 65 countries and developed economies of the world. In Finland, these students are still at the basic education level. Tables 5.1, 5.2 and 5.3 indicate that Finnish students scored at the top or are among the highest performers on the indicators of reading, mathematics and science competencies; and this pattern has been consistent for most other years.

Although these indicators may be influenced by country-specific contexts and specific trade-offs such as longer learning time as a means of conveying more and better content to students, it still reinforces the fact that Finnish students have demonstrable competencies that rank them among the best in the world of education. There have been various reasons proffered to explain the high performance of Finnish students that range from differences in state funding allocation, the competency of teachers, the student-teacher ratio and the amount of classroom hours devoted to instruction. While all these may be relevant, it still begs the question whether the exceptions would be applicable to other countries as they may have been for Finland. For instance, instead of executing a strategy of more and more, Finland has decided to educate its young people by doing less.[23]

In comparison to other countries, Lawrence Barnes[24] suggests four areas where the education policy and practice of high achieving countries (like Finland) run

Table 5.1 Finland: Comparative Results of PISA 2000

Measurement Indicators	*Score Points*	*OECD Countries Ranking*	*All Participants*
Reading Literacy	546	1st	1st
Mathematical Literacy	536	4th	4th
Scientific Literacy	538	3rd	3rd

Source: Ministry of Education and Culture 2001; OECD 2001.

Table 5.2 Finland: Comparative Results of PISA 2009

Measurement Indicators	*Score Points*	*OECD Countries Ranking*	*All Participants*
Reading Literacy	536	2nd	3rd
Mathematical Literacy	541	2nd	6th
Scientific Literacy	554	1st	2nd

Source: Ministry of Education and Culture, 2010; OECD 2012.

Table 5.3 Finland: Comparative Results of PISA 2012

Measurement Indicators	*Score Points*	*OECD Countries Ranking*	*All Participants*
Reading Literacy	524	3rd	6th
Mathematical Literacy	519	6th	12th
Scientific Literacy	545	2nd	5th

Source: Ministry of Education and Culture 2012; OECD 2013.

counter to current practice and policy in the United States: time spent in school, homework, technology and schools as agents of social change. He points out that while American students spend more time in school, do more homework and despite the increasing demand for the application of technology in the learning process, still perform lower than their comparative cohorts in countries like Finland. "In Finland, where students have scored at or near the top in comparisons of achievement for a number of years, they spend only 600 hours in school; but in the United States, the average time spent at school totals 1,100 hours, almost double that of children in Finland. Yet, American students routinely score 10–20% lower than Finnish students on international tests."[25] In 2006, while the cumulative cost per student in Finland was the equivalent of $65,000 dollars, the per student cost in the United States was approximately $95,000 dollars, but their respective student achievements based on the PISA science score was 563 for Finland and 490 for the United States.

While America leads the world in assigning homework[26]—about 140 minutes per week in mathematics—their secondary school students still post mediocre scores on math tests. On the premise that the relationship between national patterns of homework and national achievement may indicate that more homework may actually undermine national achievement results, Barnes argues that instead of focusing on setting curricula standards, specifying performance outcomes and integrating technology into the education system, there should be a broader focus in dealing with existing social problems because differences in academic achievement are more likely to be attributable to differences in social background than to variations in standardized testing.

The above conclusion may, in fact, point to some of the driving instincts with respect to the attitude of Finnish teachers both in the classroom and toward their students. As illustrated by LynNell Hancock, "Finland's improvements in reading, math and science over the past decade is due in large part to the fact that its teachers are trusted to do whatever it takes to turn young lives around."[27] The common refrain 'whatever it takes' is "an attitude that drives most of Finland's 62,000 educators in approximately 3,500 schools from Lapland in the North to Turku in the South."[28] "Many schools are small enough so that teachers know every student; and if one method fails, they consult with colleagues to try something else."[29] Teachers in Finland spend fewer hours at school each day and spend less time in the classroom than their American counterparts; hence, they use the extra time to build curricula and assess their students.[30] "In 2001, annual teaching hours of teachers in public institutions in Finland were below OECD average at primary, lower and upper secondary levels."[31]

Figure 5.4 illustrates the minimum number of weekly sessions per year and the respective subject matter focus at the basic education level. The interdisciplinary orientation of the subject matters points to the well-rounded nature and rigor of comprehensive education and the opportunity it offers to children in the development of future career interests very early in their schooling. This is particularly relevant in the sense that upon graduation from comprehensive school, students have three basic options:[32] continue education in upper general and secondary school, go on to vocational school or find employment.

Ministry of Education and Culture
Ministère de l'Education et de la culture

Subject	1	2	3	4	5	6	7	8	9	Total
Mother tongue and literature	14		14			14				42
A-language	----------		8				8			16
B-language	----------						6			6
Mathematics	6		12			14				32
Environmental Studies	Environmental and natural studies									31
Biology and Geography	9				3		7			
Physics and Chemistry					2		7			
Health Education							3			
Religion/Ethics	6					5				11
History and Social Studies	----------				3		7			10
Music	Arts and practical			4-		30	3-			56
Visual arts	Subject 26			4-			4-			
Craft, Technical Work, Textile Work				4-			7-			
Physical Education				8-			10-			
Home Economics	----------						3			3
Educational and vocat. guidance	----------						2			2
Optional subjects							(13)			13
Minimum number of hours	19	19	23	23	24	24	30	30	30	222
Voluntary A-language	----------				(6)		(6)			(12)

Figure 5.4 The Minimum Number of Lessons in Annual Weekly Sessions in Finland (Basic Education, Current at Schools).

5.4 The Finnish Way: Building Educational Value Chain

The Finnish education is a study in contrast. While it can be seen by some as unorthodox or revolutionary, it is also evolutionary in the sense that it seeks continuous improvements both in the development of curricula and in the existing mechanisms of pedagogy and delivery. On the premise that we may not know everything about the kinds of work that would be available in the future, there is therefore an existing need to continuously update the level and type of skill-sets available in society at large. The key success factor in Finland's emergence as a well-performing knowledge society with a respected education system has been its ability to reach broad consensus on most major issues concerning future directions for the country. Hence, the values already grounded in its education policy can be leveraged by combining it with emerging ICT applications to facilitate the knowledge-based society of today and for the future.

By thinking ahead and visualizing today's challenges as a microcosm of the possibilities of tomorrow, the country is able to more effectively focus its human and material resources to address emerging contingencies of the future. This is in recognition of the fact that "the environment in which skills are used is central to the value that can be derived from them";[33] hence their "economic value for society would come largely from their use in the workforce."[34] Therefore, the national premium placed on educational attainment could be said to serve three primary objectives: to develop new knowledge critical for innovation, to develop the skills necessary for employment in the workforce, and to inculcate an enduring sense of character and citizenship on the body politic. Because education in Finland is seen as a public good, the necessity for market-based competition and high-stakes testing policies has always been seen to be antithetical to the realization of the ideal of its Nordic welfare state.

Building networks and collaborative engagements among schools in different localities has remained a key to spreading and institutionalizing new knowledge and practices throughout the system. But it must start by laying the critical foundation upon which a well-structured and purposeful education system can be effected, hence the centrality of *basic education*. The following therefore provides the central distinguishing features of Finnish basic and upper general and secondary education as critical pathways of transformation that have proven crucial in overall educational performance of students in later years of schooling:[35]

- Children do not start school (Basic Education) until seven years of age.
- There is a unified curriculum for basic school instruction for all students.
- Career guidance and counseling are compulsory parts of the comprehensive school curriculum and serves as a bridge between formal education and the world of work (each student is required to spend two weeks in a selected workplace).
- The pupils rarely take examinations or do homework until they are well into their teens.
- They are not measured at all for the first six years of their education (which comprises early childhood education and pre-primary education).
- There is only one mandatory standardized test in Finland (the Matriculation Examination), taken when children are 16 years of age.
- All children, clever or not, are taught in the same classrooms.
- Schools do not select their students; rather every student is allocated a place in a nearby school.
- Finland spends around 30% less per student than the United States.
- About 30% of children receive extra help during their first nine years of school (Basic/Comprehensive Education).
- About 66% of student go on to college after basic education.
- Science classes are capped at 16 students so that they may perform practical experiments in every class.
- About 93% of Finns graduate from high school.
- About 43% of Finnish high-school students go to vocational schools.
- Elementary school students get 75 minutes of recess a day in Finland, versus an average of 27 minutes in the United States.
- Teachers only spend four hours a day in the classroom, and take two hours a week for *professional development* training.
- Finland has the same amount of teachers as New York City, but far fewer students.
- The school system is 100% state-funded, although some private contributions do exist.
- All teachers in Finland must have a Master's degree, which is fully subsidized by the state.

- Teachers are well respected, relatively well paid and have greater autonomy and accountability.
- The national education curriculum is only based on broad guidelines; teachers assess their pupils based on these guidelines; from this assessment they will be selected for further studies in general and upper secondary school or vocational education.
- Teachers are selected from the top 10% of graduates.
- High school teachers with 15 years of experience make 102% of what other college graduates make.
- There is no merit pay for teachers.
- Teachers are given the same professional status as doctors and lawyers.
- There is a strong focus on self-evaluation of schools and education providers, but national evaluation of learning outcomes are done regularly.
- Responsibility for the operations of basic educations schools and upper secondary schools rests with the principals.
- Based on an international standardized measurement in 2001, Finnish children came in at the top, or very close to the top, in science, reading and mathematics.

But still, the Finnish education system is an evolving project, one that is continuously seeking new means of engagement and renewal. By embracing the principle of self-evaluation and accountability as a means of detecting potential weaknesses, whether in personnel, training or the existing structure of pedagogy, the system is able to respond effectively to variations from its own in-built signals. The collective sentiment has always been that just because something is working well at the moment does not mean that we cannot find other ways to make it even better. While the moral behind all innovations is the need to find new and more effective ways of doing things, it is important to note that innovation is also a process of learning—learning from experience or learning by thinking outside the box. Table 5.4 illustrates five pedagogical pathways and the essential features that make up the Finnish Way—a pragmatic deviation from old notions of and approaches to the teaching and learning experience.

Education is not only about developing mastery in a subject matter; real education is about being educated about everything else that may impact on the subject

Table 5.4 The Finnish Way: New Model for Global Education Reform

Traditional Way (Old)	*The Finnish Way (New)*
Core subject	Holistic learning
Standardization	Personalization
Competition	Collaboration
Choice	Equity
Control	Trust

Source: Compiled by author based on information from Pasi Sahlberg, *Finnish Lessons: What can the world learn from Finland*? Presentation at the Annual Leadership Conference, Birmingham, England, June 16, 2011.

matter, directly or indirectly. The essence of holistic learning, therefore, is that it offers a means of building an interdisciplinary body of knowledge that, in the end, could provide multiple pathways for addressing various social and economic needs. While a personalized learning process is a means of giving close attention to individual student needs and capabilities, collaboration is essential for building the networking critical for co-creativity and innovation. In the sense that equity can help to reinforce the principle of equality and fairness, trust grants legitimacy as a function of the perceived appropriateness of the logic and order in which things are done. In theory, and seemingly so in practice, these have essentially been the guiding norms of Finland's education policy.

Notes

1 Suvikumpu, L. 2010. *Find out about Finland*. Keuruu, Finland: Otava Publishing Co. Limited, p. 26.
2 Taylor, A. 2012. Why Finland's unorthodox education system is the best in the world. *Business Insider*, p. 1. Historically, public education in Finland began in the 1860s as a result of the influence of the Lutheran Church. A national school system independent of the church began in 1866, and three years later, a Supervisory Board of Education was established under the Ministry of Education to inspect, monitor, and govern the school system in Finland.
3 Ministry of Education and Culture. 2013. The Finnish education and decision making system, p. 2.
4 OECD. 2013. Governance: Autonomous decision making at the local level, *Education Policy Outlook: Finland* www.oecd.org/policyoutlook.htm, p. 14.
5 Ministry of Education and Culture. The Finnish education and decision making system, p. 9.
6 Sahlberg, P. 2011. Finnish lessons: What can the world learn from educational change in Finland? New York, NY: Columbia University, p. 15.
7 Korhonen-Yrjanheikki, K. 2007. New challenging approaches to engineering education: Enhancing university-industry co-operation. *European Journal of Engineering Education*, 32(2): 167–179, p. 168.
8 The Europe 2020 Strategy can be located at: http://ec.europe.eu/europe2020/pdf/csr2014/nrp2014_finland_en.pdf, Accessed on July 29, 2015.
9 CIMO. 2014. Educational authorities in Finland. www.studyinfinland.fi/destination_finland/education_system/educational_authorities. Accessed on July 29, 2015.
10 *Economist, The*. 2015. Excellence v equity, Special Report on Universities, March 28, p. 4.
11 Ibid.
12 Kalu, K. N. 2011. Institution-building, not nation-building: A structural-functional model, *International Review of Administrative Sciences*, 77(1): 119–137.
13 Parsons, T. 1937. The structure of social action. New York, NY: McGraw Hill. Also see Talcott P. 1956. A sociological approach to the theory of organizations. In *Structure and processes in modern societies*, ed. T. Parsons, p. 27. Glencoe, IL: Free Press.
14 *Economist, The*. 2015. An hereditary meritocracy. January 24, p. 18.
15 Hancock, L. 2011. Why are Finland's schools successful? *Smithsonian Magazine*, September, p. 3.
16 OECD. 2013. Equality and quality: Ensuring educational opportunity for all, *Education Policy Outlook: Finland* www.oecd.org/policyoutlook.htm, p. 6.
17 OECD. 2013. Preparing students for the future: Well-developed general and vocational education and training programmes, *Education Policy Outlook: Finland* www.oecd.org/policyoutlook.htm, p. 8.

18 Sahlberg, P. Finnish lessons: What can the world learn from educational change in Finland? New York, NY: Columbia University, p. 32.
19 OECD. 2013. Preparing students for the future: Well-developed general and vocational education and training programmes, p. 8.
20 Hancock, L. Why are Finland's schools successful?, p. 2.
21 OECD. 2013. Preparing students for the future: Well-developed general and vocational education and training programmes, p. 8.
22 This paragraph is an excerpt from OECD. 2013. School Improvement: Trust and proficiency of school leaders and teachers. *Education Policy Outlook: Finland* www.oecd.org/policyoutlook.htm, p. 10.
23 Baines, L. 2007. Learning from the world: Achieving more by doing less, *Phi Delta Kappan*, 89(2): 98–100.
24 Baines, L . Learning from the world: Achieving more by doing less, p. 23.
25 Ibid. p. 24.
26 Ibid. pp. 24–26.
27 Hancock, L. Why are Finland's schools successful?, p. 1.
28 Ibid.
29 Ibid.
30 Ibid. p. 3.
31 OECD. 2013. School improvement: Trust and proficiency of school leaders and teachers, *Education Policy Outlook: Finland* www.oecd.org/policyoutlook.htm, p. 10.
32 Sahlberg, P. Finnish lessons: What can the world learn from educational change in Finland? New York, NY: Columbia University, p. 23.
33 Economic Intelligence Unit. 2014. The learning curve: Education and skills for life 2014 report, New York, NY: Pearson, p. 13.
34 Ibid. p. 11.
35 Compiled by author based on information drawn from A. Taylor. 2012. Why Finland's unorthodox education system is the best in the world. *Business Insider*. URL: http://www.businessinsider.com/finlands-education-system-best-in-world-2012-11?op=1#ixzz3aADmiNAq.

6 Health and Social Policy

Nordic Welfarism

In the book aptly titled *Why Are Some People Healthy and Others Are Not?*, Hertzman et al. argue that the health status of populations is determined by the interaction of various factors such as "the features of the social and physical environment, the genetic endowment, and the learned biological and behavioral responses of individuals, and access to and responses of the health care system."[1] While this is not meant to be exhaustive, suffice it to say that a society's health status is a function of the social climate, health-related behavioral and biological risk factors, general susceptibility, psychosocial factors, and the condition and characteristic of the physical environment. In the sense that "social climate shapes the behaviors of its individual, who in turn influence the environment,"[2] it would be difficult to determine every direct causal pattern in the occurrence of all major disease categories. Hence, health policy is often made by way of generalizations based on hypothetical causal pathways or past precedents.

For most states, it is a question of dealing with what you already know as well as what you do not know, but in the final analysis, it often comes down to the level of preparedness in terms of national health infrastructure, the number of physicians per 1000 population, the level of skill and specializations at hand and the health care budget as a percentage of GDP as well as per capita. Because health policy cannot be made in a vacuum, states would need to develop the requisite capacity to deal with existing and contingent health issues but also have the resources to sustain them for the long haul. The high standard of living and the spate of rising expectations and changing lifestyles in many western industrialized societies mean that governments must develop the requisite capacity to meet current and future social challenges.

In response to these challenges, the Nordic countries have embraced the concept of the welfare state as a means of addressing issues of equality, economic, social and environmental sustainability. On the belief that everyone has a "right to social well-being, political participation, and equitable healthcare, it therefore becomes the state's responsibility to create a functioning social protection system that helps citizens cope with the risks and challenges of life."[3] The Finnish social welfare system covers the whole life cycle from pregnancy (maternal and child health) to death (including funeral allowances), and the employer is bound by law to ensure health and safety at work.[4] With the simple reality that not everyone can enjoy all benefits and privileges of citizenship to the same quantity, quality and degree, the welfare state therefore aspires toward a normative as well as a balanced redistributive end.

6.1 On the Principles of the Welfare State

Finland has a universal comprehensive welfare system that covers everybody rich and poor, privileged or not—made up of a national flat-rate pension available to all once they meet the age criteria (whether they paid into the plan or not) and a national health plan determined by medical and health needs rather than ability to pay. In the same way the United States grants (social security payments, Medicare, veterans' benefits, etc.) as entitlements, but in Finland, there is a legal guarantee that imposes on the government the legal obligation to secure the collective well-being of the citizens at large. In the same way that the welfare system guarantees collective benefits, it also demands collective costs and sacrifices for its sustenance, albeit differentially. The revenue needed to sustain the welfare system comes largely from the state, employers and employees tax revenues.

In fundamental ways, particularly in institutional structure, financing and objectives, Finland's health care system resembles those of most other Nordic and EU countries "in that it covers the whole population and its services are mainly produced by the public sector and financed through general taxation."[5] The health care system is much decentralized through several municipal and district-based hospital systems. Like most EU countries, Finland is confronted with an aging population and rising public expectations, but most important is "the continuing challenge in delivering equal, efficient and high quality health services at affordable costs in times when the amount of care needed seems to be exceeding the resource base for providing it."[6]

As a result, "the health system faces the problem of rationing services in order to cut costs owing to a simultaneously increasing demand for health services and a decreasing tax base to pay for it."[7] Although the decentralized health system has increased public access, it has also created regional disparities between wealthier and poorer municipalities, making it more difficult to use the same set of indicators to measure national health outcomes. Today, many municipal health systems are facing a series of problems, among them increasing tax and financial pressures, changes in migration and population trends leading to changes in the types of and demands for health services, an aging population and other structural factors associated with labor and specialized medical training for the treatment of target populations.

Because the social welfare system guarantees basic universal and legislative rights, the state's obligation to offer social protection to all of its citizens (rich or poor) is anchored on the premise of health care as a public good—and thus "ensure universal access as well as a greater role for the government in the health care system."[8] There are several important milestones in the evolution of the Finnish welfare system as far as it concerns public health, both in service delivery and financing. First was the passage of the Hospital Act (1956), then the Sickness Insurance Act (1963), the Primary Health Care Act (1972), the Occupational Health Act (1978), the Act on Specialized Medical Care (1989), the Health for All by Year 2000 Programme, Universal Dental Coverage (2002) and consequent reform measures that culminated in the Health Care Act (2011). Because of the health-related "legislations passed in the 1950s and 1960s that mandated the construction of a network of hospitals, and the education of more medical personnel; policy makers began to emphasize the use of smaller clinics from 1963 to the early 1970s," as a means of extending more effective health care and opportunities to those at the rural areas.[9]

While the Health Care Act of 1972 helped to decentralize health care facilities from the mostly urban centers to local communities, increase access and reduce health inequalities between different population groups and grant more support for out-patient care, the Health Care Act of 2011, on the other hand, sought to improve freedom of choice for patients and the quality of care and to bring primary health care units under the control of the various hospital districts. Furthermore, the earlier passage of the Social Welfare Act (1982)[10] made the general planning, guidance and supervision of social welfare issues the responsibility of the Ministry of Social Affairs and Health, particularly as it concerns social services, social assistance and allowances, loans and related measures meant to promote and maintain the social security and functional capacity of the individual and community.

As a result, the country has managed over the years to achieve very dramatic scores on various global health indicators. The country has been able to reduce the prevalence of tuberculosis cases from a high of 1,400 notified cases in 1945 to a negligible low of about 250 cases in 2005; age-standardized chronic heart disease mortality in men and women aged 25 to 64 years declined from 390/1000 cases in 1952 to about 50/1000 cases in 2009. Average life expectancy has increased to 77 years and 84 years for men and women, respectively; and communicable diseases are under control. Based on the 2014 CIA *World Factbook*[11] country profile, Finland's infant mortality rate was 2.52 deaths/1000 live births, total life expectancy at birth is 80.77 years, 2.91 physicians/1000 population (2009), 5.5 beds/1000 population (2011), obesity prevalence rate was 22.8% of the adult population, death rate was 9.8 deaths/1000 population (2015).

With a total health care expenditure of about 9.4% of GDP (2013 estimates), and a total population dependency ratio of 56.8%, the cost of sustaining the healthcare system would continue to rise for the foreseeable future, and this considered in light of the aging population and the rise in the elderly dependency ratio to the tune of 31% of the population. In the current age structure, those 55 to 64 years old make up approximately 13.56% of the population, and those 65 and over make up about 20.21%, totaling 33.77% of the population. If we include the fact that people aged 25 to 54 years currently account for about 38.03% of the population, then it would not be too long before they too start to place further strains on the financial and infrastructural support needed to sustain the health care and social welfare system. One of the challenging issues that has been raised regarding the Nordic social welfare and health care systems is the concern for long-term sustainability in an era of fiscal stress, economic downturn, rising equipment costs as a result of the fast-paced nature of technological obsolescence in the health care economy and the increasing demand premised on quality of life and maintenance factors. They face the same situation as in the United States; the people want better quality health care, but they are not willing to raise taxes to pay for it, even in the face of evident realities that may speak to the contrary.

6.1.1 Policy and Standards of Practice

The Finnish government sets the broad national priorities and objectives of health care and proposes them to Parliament as a legislative bill to be deliberated and possibly passed into policy. As part of the government, the Ministry of Social Affairs and Health directs and guides policy development related to health care and social welfare within stipulations set in the Government Programme, prepares legislations and key reforms (as applicable), steers the implementation process and engages the various actors and players involved with the political decision-making process.[12] The next layer is the provincial state offices,[13] which promote the national and regional objectives of the central administration and keep contacts with municipalities in their jurisdictions. The social and health departments of the provincial administrations are responsible, among other things, for guiding and supervising both municipal and private health care providers. There are five regional provinces, plus the autonomous Åland Islands. The municipal authorities exercise a high degree of sovereign power in policy making in the areas of health care and social welfare. Since they have been granted power by law to provide these services in their local communities, municipalities are able to levy taxes as well as user fees, complemented by state subsidies as a means of generating the needed revenue.

The main decision-making power in the municipalities is the municipal council,[14] which is elected every four years by the inhabitants of the municipality. Decisions on the planning and organization of health care are made by the health committee (delegated by the Municipal Council), the municipal council and the municipal executive board. Also, the hospital district administration exercises power over the organization of specialist hospital services within the municipal health care system and approves financial guidelines such as activity-based payment systems, rate or price structures, and decisions about key investment choices. The views of municipal residents, physicians and other health care providers, pharmacies and the general national public opinion can, in fundamental ways, shape the outcome of policy deliberations and health care choices that are made within municipalities. Hence, in matters of policy and decision making, no single actor is dominant. There is always a "constant balancing act between the roles of the public and private sectors in all activities related to the health care system, from controlling costs to the allocating resources,"[15] constituting "what might be construed as a fluid policy community with shifting alliances, assumptions, and coalitions of interests."[16]

6.2 The Structure and Organization of the Health Care System

As alluded to in earlier sections of this chapter, the Finnish health care infrastructure and delivery system is highly decentralized with municipal and local authorities exercising a reasonable level of autonomy in decision making, financing and the allocation or re-allocation of resources to better serve their respective constituencies. Because "the goal of health care in Finland is to maintain and improve mental health and physical functional capacity through preventive and comprehensive health services,"[17] there are three different health care systems in the country that receive public funding: "municipal health care, private health care, and occupational health care, with significant differences in the scope of the services provided, user-fees, and waiting times."[18]

Municipal Health Care

The municipalities are responsible for providing primary health care coverage for all those who live and are permanent residents of their jurisdictions. "Primary health services provided by municipalities are defined in the Primary Health Care Act of 1972 which authorizes that every municipality must have a health center which provides primary services, either independently or by creating joint municipal health centers."[19] Primary health care[20] refers to first-contact care, in which the majority of health problems are treated. It is the foundation of any health care system, and nations with strong primary care seem to have better health outcomes than those without it. As noted in the World Health Organization's Declaration of Alma-Ata, 1978, primary health care is a broader concept that in addition to primary care services includes health promotion and disease prevention, and also population-level public health functions.[21] A sizeable proportion of the services rendered by the primary health care centers are out-patient visits, offering dental care, educational and counselling services, but also advice on preventive medicine and better management of individual patient health status. Specialized in-patient care[22] is provided through hospital districts that are managed and funded by the municipalities under the Specialized Medical Care Act of 1989.

Although each municipality must be a member of one hospital district, each hospital district has a central hospital, five of which are university-level teaching hospitals (Figure 6.1). Except in cases of life-threatening emergencies, a referral from a licensed physician is needed for access to medical care offered

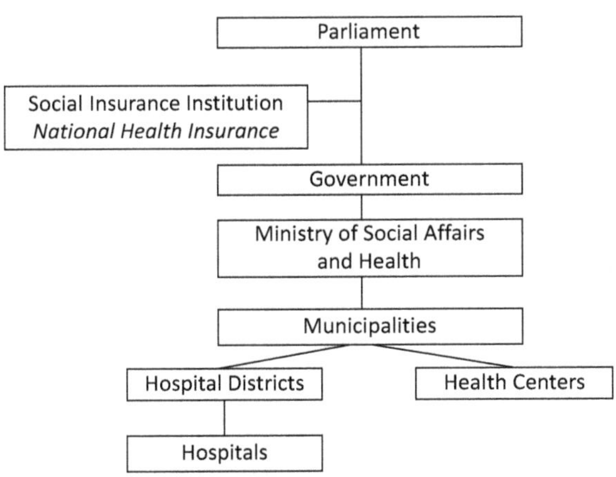

Figure 6.1 The Structure of Finland's Health Care System.

Source: Saarivita, Toni, Davide Consoli, and Pieter Dhondt, 2012.[25]

at the hospital districts. "Municipalities can also purchase health care services (primary health services or specialized health care services) from other municipalities, private providers, other hospital districts, or non-governmental organization (NGOS) and foundations."[23] There are about 336 municipalities as of 2012, down from a high of about 415 in 2008: five university hospitals and about 20 or more hospital districts. This does not include the Åland Islands,[24] an autonomous Swedish-speaking part of Finland with 16 municipalities and approximately 26,000 inhabitants. The Åland government is responsible for providing healthcare services in the region, but they can purchase from either Sweden or Finland health services that are not readily available within the region.

As illustrated in Figure 6. 2, and based on physician referral, a patient's treatment protocol can be migrated from one hospital district system to another, based on the type of treatment regimen and medical specialization required. In hospital X, visits 1 and 2 would ordinarily involve out-patient services, but if the required specialized care is not available in hospital X, then a referral to hospitals Y and Z would be needed to complete the total episode of care. An episode of care can also be completed within one single hospital district. A referral that would involve the whole chain of three hospitals (including specialist care) would remove the incentives for physicians or providers to skimp on care because the price system is already established through existing or contractual Diagnosis Related Groups (DRGs) schedules—a patient classification system that allows hospitals and providers to group or bundle disease categories based on their relatedness and scope of resource consumption (including time and money). It is generally seen as a cost-saving and control tool that allows health subscribers (insurance companies, institutions or organizations) to pay physicians, hospital

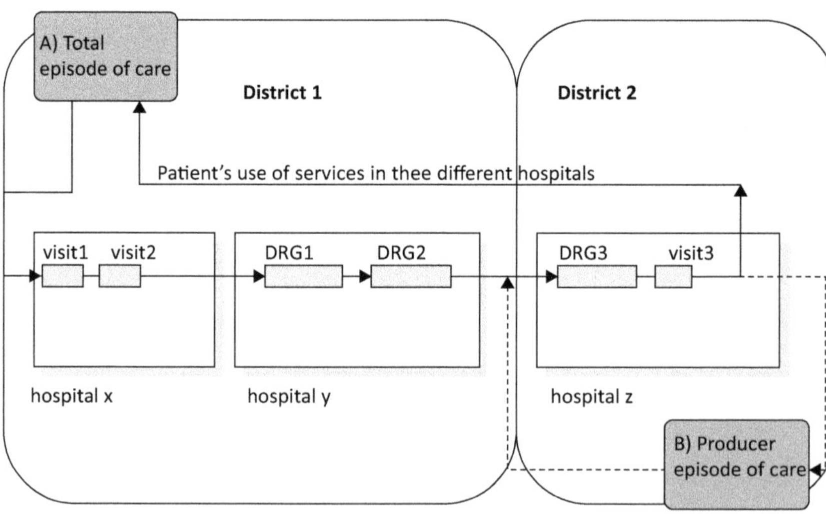

Figure 6.2 A Functional Approach to Referral and Total Episode of Care.
Source: Unto Häkkinen 2010, p. 17.

systems, or other medical providers a discounted amount for the whole episode of care and related basic or post-acute services. In Finland, "two different DRG systems are used for health service costing and invoicing: the NordDRG Classic for in-patient treatment and day surgery; while the NordDRG Full System is used for in-patient and out-patient treatments at hospitals."[26]

Private Health Care

There are various private hospitals that provide about 5% of total health care demand. Approximately 25–30% of specialized out-patient care is conducted by private-sector hospitals, partly due to their ability to expand nationwide, as well as the variety of specialized care they can offer to municipal health centers or hospital districts. While some of them are for-profit hospitals working on a fee-for-service basis, patients can either pay from out-of-pocket or through physician referrals linked to an employer-based insurance scheme. There are also non-for-profit hospitals that provide general or specialist health care. Municipal authorities can also contract services from some of the specialist private hospitals, especially in such areas as orthopaedic medicine and neurological and related services.

Occupational Health

Finland's "first legislation on occupational safety and health was passed in 1889, and the development of occupational health serviced (OHS) infrastructure started immediately following industrialization."[27] Driven by an overall emphasis on improving working conditions and lengthening working life in general through a higher standard of occupational health, safety, risk prevention and well-being at work, the consequent passage of the Occupational Health Care Act (1978) "aimed to increase preventive measures focusing on work and working conditions and enhancing the scope of occupational health services (OHS) as a continuous process that includes workplace monitoring and assessment, operational planning, quality improvement, and medical check-ups."[28] The Department of Occupational Safety and Health plays a leadership role in making policy regarding occupational health issues under the guidance of the Ministry of Social Affairs and Health.

The Act made mandatory for all employers who have at least one salaried employee in the company, but voluntary for self-employed entrepreneurs,[29] that services offered under occupational health care include "physical examinations and first aid for employees at the work place, as well as periodically checking the health status of all employees whose work might endanger their health."[30] Occupational health services (OHS) are generally offered to companies and entrepreneurs located in each municipality, either from work-related episodes or for curative out-patient services. Although the national health insurance (NHI) scheme covers about 40% of the expenses,[31] the employer is entitled to receive compensation for necessary and reasonable costs incurred through occupational health services. While the "Social Insurance Institution (SSI) pays for this compensation out of earned income insurance for which contributions are collected from employers (73%) and employees

(27%),"[32] the operational cost structure is determined based on prorated allocations assigned to specific health or treatment conditions. A population survey indicates that "about 45% of physician visits of employed people were in occupational health care, 35% in municipal health care and 15% in private health care."[33] While "there are significant differences between the systems (i.e. in terms of scope of services, user-fees, and waiting times), the more readily available choice for low-income people has been the municipal health care system."[34]

Pharmacies

Generally, pharmacies[35] are privately owned and are heavily regulated by the government. In addition to the fact that their margins and prices are fixed by the government, they cannot be owned by companies or conglomerates. The National Agency of Medicines (NAM) decides in which locations pharmacies will be located and chooses the pharmacist to run them. While out-patient drugs are partly reimbursed under the national health insurance scheme, they are paid directly to the pharmacies. Other types of care and utilization within the Finnish health care delivery system include mental health, maternal and child health and geriatric care, as well as *emergency care* for all including immigrants and asylum seekers. Generally, "asylum seekers, illegal immigrants, tourists, temporary students and workers (from non-EU countries) are not covered under municipal health care; however, as per the *Act on Adaptation of Immigrants and Reception of Asylum Seekers*, those seeking formal asylum in Finland are entitled to the same health services as permanent residents."[36] The health care services to persons from other EU countries are provided at the same benefit level, but the costs have to be paid out-of-pocket or through facilities from their home countries.

To improve the effectiveness of the delivery systems and to create more access for the various client groups, consequent efforts at reform have focused on eliminating redundancies in services (especially in those areas where occupational health care and primary health care intersect), "to reduce inequalities among the regions and among population groups as well as to monitor expenditure streams associated with undesirable and unnecessary use of critical health care services."[37] There has also been increased used of case-based pricing schemes in the form of pre-negotiated capitation rates between municipal authorities/national health insurance and private providers that serve the municipal or district-based hospital systems. This has helped, in some ways, to contain rising health care costs.

6.3 Financing Health Care and Its Challenges

Broadly speaking, there are two sources of public financing for health care services: municipal financing based on municipal taxes, state subsidies and user-fees and national health insurance (NHI) based on mandatory insurance fees.[38] The average share of total pubic health care expenditures borne by "municipal authorities is 33%; about 29% comes from state subsidies (mainly raised from

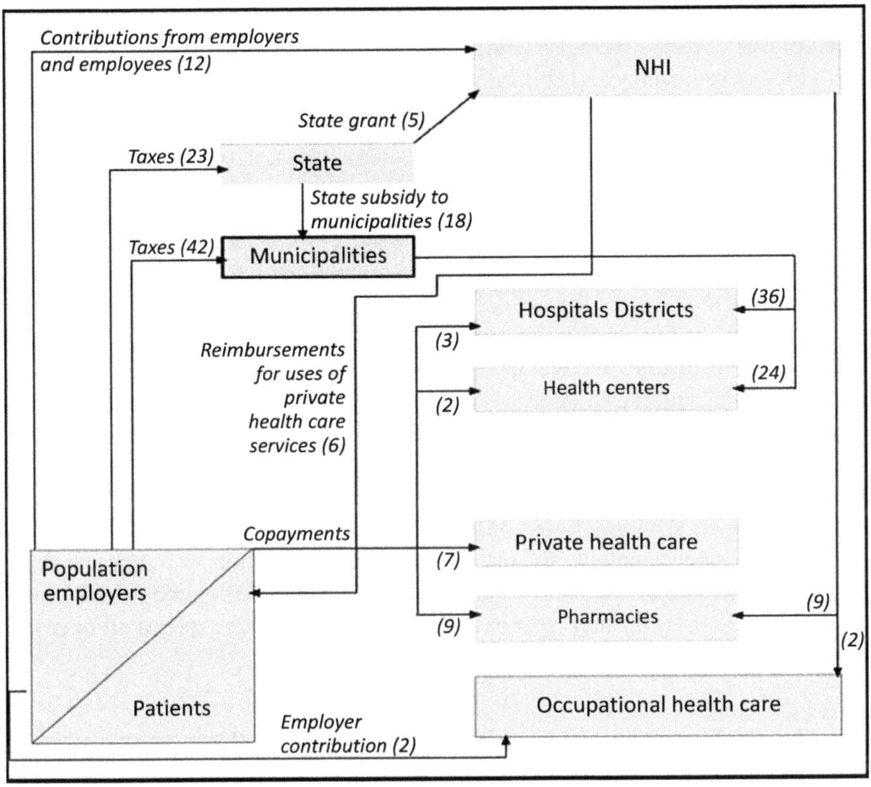

Figure 6.3 Hospital and Health Care Funding Flows in Finland.
Source: Adapted from Unto Häkkinen, 2010, p. 3.

general tax revenues), 13% from the national health insurance scheme, and while private insurance contributes about 2%, the rest comes from private out-of-pocket sources and co-payments by patients"[39] (See Figure 6.3). In almost all cases, municipalities fund as well as own the health service provision organizations (health centers and hospital districts) in their areas, but they do not fund out-patient drugs and transportation costs. On the other hand, the national health insurance scheme provides compulsory insurance to all residents of Finland and "funds private health care, occupational health care, outpatient drug and transportation costs, and sickness allowance—through sickness insurance and income insurance."[40]

The coverage under sickness insurance includes outpatient drug reimbursement, partial reimbursement of private sector and rehabilitation services, ambulance services and transportation costs to health care centers; while income insurance covers sickness allowance, maternity leave and rehabilitation allowances, as well as reimbursement to employers for occupational health services rendered to employees (excluding dependents). Funding for sickness insurance comes from employee deductions and subsidies from the state, while income insurance is funded by employees, as well as a percentage of the earnings paid by the employer. Other coverages within the NHI scheme include "loss of income due to illness, pregnancy and childbirth, and loss of income of the parents of a sick child during treatment and rehabilitation of the child."[41]

Private Health Insurance

Private health insurance is not very widely used in Finland, partly as a result of rising costs, ability to pay, and the risk-rated nature in the estimation of premiums, but also due to the fact that the publicly financed system and the role of municipalities in overall health care delivery means that there are multiple options to secure access to health insurance coverage other than the private option. Private health insurance comes in three categories:[42] voluntary health insurance, statutory motor accident insurance and statutory occupational accident insurance (compulsory for every employer). "The total share of the Finnish population that has private health insurance is about 11%."[43] One of the most common observations has been that "most individuals who purchase voluntary private health insurance do so to limit their out-of-pocket payments for private care and the portion of outpatient drug charges not covered by the NHI."[44] The NHI is administered by the Social Insurance Institution (SII) through approximately 260 local offices spread all over the country.

6.3.1 Impact Assessment: Critical Indicators and Renewal

It would be acceptable to make the case that every health care scheme comes with its own problems and peculiarities. Perhaps what works in one geopolitical context, might be quite problematic in others. But one common denominator among all schemes is how to maintain access, reduce cost, and enhance quality. For the Finns, these seem to present a moral as well as a constitutional imperative. How to sustain state subsidies in the face of an aging population, rising costs tied to technological obsolescence and development in the health care sector? How to maintain a sustainable funding stream for the social welfare system in the context of a declining contribution base and rising public demand as a result of rising expectations and improvements in the quality of life? While "investing in technology and changing the way care is delivered is complex, expensive, and takes time to implement effectively,"[45] it still begs the question whether such investments always balance out in terms of increased access, lower costs, and better overall health outcomes.

"The state subsidy to municipal social welfare and health care expenditures is determined by the population, age structure and morbidity in the municipality plus a number of other unique factors. The subsidies constitute about 25%–30% of municipal spending on health services."[46] In 1993, a major reform of the state subsidy program occurred, in which "subsidies for determining costs in municipal health services are non-earmarked lump-sum grants, but are calculated prospectively by using a specific need-based capitation formula."[47] "Particularly in the field of specialist hospital care, the reform meant that the system changed somewhat from a public integrated model to that of a pubic contract model."[48] This is the essence of managed care (in the US context) in which a "capitation-based contract model is used as a method of physician payment along with other providers such as general practitioners as gatekeepers"[49] to reduce cost and improve quality.

Under existing disease management protocols, specific governance mechanisms are included in the contract to prevent physicians or providers from either *cherry-picking* (selecting the healthier and less expensive patients while avoiding patients whose treatment would take longer and cost more money) or *skimping* on care (delaying or denial of services to members of a prepaid or "capitated" health plan or reducing approvals of costly procedures) as a means of controlling costs and increasing profits. However, in light of the provisions of the Health Care Act of 2011 which now grants the Finnish population increased freedom and rights to choose where and who would be responsible for providing them with needed health care services; how would this affect the sustainability of the state subsidy program for the future? While a retraction in the scale of the state subsidies would mean an increase in the cost of health care for the primary care offered at municipal health centers and the district hospitals, it could also generate a long-term trend of rising deductibles as more people become discouraged from seeking appropriate care. This would, invariably, lead to a general decline in national health outcomes and quality.

Considered in light of the current economic realities in Finland, how realistic and dire could this be as a possibility? As of 2014,[50] Finland had a current budget deficit of -3.5% of GDP and a public debt of equal to 59.6% of GDP (or $589.9 billion based on 2012 estimates); it generated revenue of $146.3 billion and an expenditure of $156.1 billion, yielding a revenue shortfall of $9.8 billion. With 2014 total exports amounting to $78 billion dollars and total imports of $73 billion, Finland had a surplus of about $5 billion. But then the trade surplus represents a paltry sum when considered along with the existing public debt of $589.9 billion dollars and a negative budget deficit. Since health care provision is a constitutional obligation of the state, it represents a public entitlement and hence must be honored even if the state should go into additional debt to do so.

On the one hand, the ability to maintain the subsidy, at least at the current rate, would depend on the state of economic growth, most particularly in the export sector. Because the export sector is dependent on the international market demand for Finnish products and technology, its fate would be, more or less, influenced by the uncertainties and vagaries of the international market regime. Perhaps an alternative would be to peg the rate of growth and decline in the allocation of state health care subsidies in accordance with reciprocal changes in the rate of economic growth. A declining export economy would mean that in order to sustain the state subsidies and the social welfare system that goes with it, the state must go into further debt to keep the publicly funded health care system alive. But for how long and at what cost?

6.4 Preventive Medicine: Health Promotion and Nutrition

The issue of health promotion has remained a central element of Finnish health care policy and is funded at both the national and municipal levels. The primary institutions in the administration of this sector are the Ministry of Social Affairs and Health and the respective municipal authorities. The rationale behind health promotion has been part of an increasing emphasis on the need to shift the focus from treating the sick to actively promoting well-being; "developing performance indicators based on specified objectives, the use of mass media to publicize health-enhancing behaviors and attitudes, and a focus on working with communities to develop health-enhancing environments."[51] In the end, it is expected that this would help "to improve quality of life, increase individual functional ability and boost human productivity while restraining the growth of social expenditure on health."[52] The Health Promotion Act of 1994 was aimed at promoting individual and general population health and to prevent or reduce such factors in the living environment that may constitute a health hazard. In this regard health promotion involves not only individual but also environmental health, mental health, physical fitness, control of disease bearing pathogens, pollutants, infectious diseases and airborne spores.

One of the essential objectives "of health promotion is to empower people to take increased control over aspects of their lives that could affect their own health and the health of others"[53] and "in doing so, help to reduce the financial burden on health care services"[54] across all aspects of the health delivery continuum. Hence, investment in health promotion is a mutual responsibility that leads to mutual gain for individuals, public organizations, private businesses and society as a whole. In all advanced economies, it has been expanded to include exercise and fitness regimes and nutrition and healthy eating habits; health education to prevent the occurrence of diseases, obesity, drug and substance abuse; cessation or reduction of smoking, alcoholism; dietary changes to reduce fat consumption; and a decrease in specific risk behaviors and lifestyle choices.

Finland's approach to health promotion utilizes an intersectoral and multi-agency approach and by working with various government agencies, non-governmental organizations, units of the World Health Organization, the European Union and pan-Nordic committees, it is able to provide a coordinating mechanism similar to the one found in most Nordic countries. While issues of food safety are handled by the Ministry of Agriculture and Forestry and the Finnish Customs, nutrition and healthy eating habits are also handled through initiatives of the Nutrition Committee sanctioned by the Ministry of Social Affairs and Health; anti-smoking programs are also shaped partly through the tax policies of the Ministry of Finance.

6.4.1 Value-Based Health Care: Quality and Cost

Finland's health care system is ranked among the best in the world, which in itself, does not necessarily mean that it does not face its own peculiar challenges. Some of those are the increasing regional disparities in access and health outcomes, rising costs due partly to budgetary constraints and also the increasing application of technology and innovation in the health care sector.

Because municipal health centers and hospital districts have come to be the operational backbone of the Finnish health care system, there should be a renewed focus on a structural reform of the delivery mechanisms and benchmarking to determine what and where improvements are needed across the existing continuum of care.

In addition to networking, a more enhanced form of horizontal integration across and between functional units would enable poorer municipal health centers to leverage the capabilities and assets of the more disposed municipalities and hospital districts. In the end, the most crucial question would be one that seeks ways to "achieve optimal health outcomes for the Finnish people (including reductions in regional disparities) in a financially and socially sustainable way."[55] Entitlements of any kind including social welfare and health care programs are difficult to manage without going into obvious debt, most particularly since they do not lend themselves easily to the adjustments necessary to accommodate changing national economic fortunes. The political imperatives they generate, more often than not, seem to outweigh the more practical issues of revenue adjustments and affordability. But the Finns seem not to have put a price on the moral reasoning justifying its Nordic-style welfarism.

Notes

1 Hertzman, C., J. Frank, and R. C. Evans. 1994. Heterogenetics in health status and the determinants of population health. In *Why are some people healthy and others not*?, ed. R. G. Evans, M. L. Barer, and T. R. Marmor, 67–92. New York, NY: Aldine De Gruyter, p. 68. Also see R. C. Evans and G. L. Stoddard, *Producing health, consuming health*, pp. 27–64.

2 Patrick, D. L., and T. M. Wickizer. 1995. Community health. In *Society and health*, ed. B. C. Amick III, S. Levine, A. R. Tarlov, and D. C. Walsh, 46–92. New York, NY: Oxford University Press, p. 60.

3 Ministry of Social Affairs and Health. 2010. Toward social sustainability, *Social Sustainable Finland 2020: Strategy for Social and Health Policy*, December, p. 4.

4 World Health Organization. 2012. National profile of occupational health system in Finland, Copenhagen, Denmark: Regional Office for Europe, p. x.

5 Häkkinen, U. 2010. Financing of hospital care in Finland. *Euro Observer 12* (3): 10–12. https://www.google.com/?gws_rd=ssl#q=unto+Hakkinen%2C+Financing+of+hospital+care+in+Finland, p. 10. Accessed on August 15, 2015.

6 Jakubowski, E. and R. Busse. 1998. Health care system in the EU: A Comparative study. Luxembourg: Directorate General for Research, European Parliament, p. 5.

7 Ibid.

8 Graig, L. A. 1999. Health of nations: An international perspective on U. S. health care reform. Washington D. C: CQ Press, p. 179.

9 www.photius.com/countries/finland/society/finland_society_income_security_prog_13.html Accessed on August 1, 2015, p. 2.

10 See Social Welfare Act 710/1982, Ministry of Social Affairs and Health, Finland.

11 CIA World Factbook. http://www.cia.gov/library/publications/the-world-factbook/geos/fi.html. Accessed on August 8, 2015

12 Doupi, P, E. Renko, P. Hämäläinen, M. Mäkelä, S. Geist, and J. Dumortier. 2010. E-health strategies country brief: Finland. Brussels, Belgium: European Commission, p. 11.
13 Vuorenkoski, L., P. Mladovsky, and E. Mossialos. 2008. Finland: Health systems review, *Health Systems in Transition* 10(4): 1–168, p. 4.
14 Doupi, P. et al. E-health strategies country brief: Finland, pp. 11–12.
15 Graig, L. A. Health of nations: An international perspective on U. S. health care reform, p. 3.
16 Kalu, K. N. 2005. Competing ideals and the public agenda in Medicare reform: The 'garbage can' model revisited, *Administration & Society* 37(1): 23–56, p. 30.
17 World Health Organization. 2012. *National profile of occupational health system in Finland*, Copenhagen, Denmark: Regional Office for Europe, p. x.
18 Doupi, P. et al. E-health strategies country brief: Finland, p. 9.
19 Ibid. p. 12.
20 http://www.med.uottawa.ca/sim/data/Primary_Care.htm. Accessed on August 5, 2015.
21 http://www.euro.who.int/__data/assets/pdf_file/0009/113877/E93944.pdf?ua=1, Accessed on August 5, 2015.
22 Doupi, P. et al. 2010. E-health strategies country brief: Finland, p. 12.
23 Vuorenkoski, L., P. Mladovsky, and E. Mossialos. Finland: Health systems review, p. 3.
24 Ibid.
25 Saarivirta, T., D. Consoli, and P. Dhondt. 2012. The evolution of the Finnish health-care system early 19th century and onwards. *International Journal of Business and Social Science* 36(6): 243–257. Also see Järvelin, J. 2002.Health care system in transition. *European Observatory on Health Care Systems*, 17–20; and Häkkinen, U. 2005. The impact of changes in Finland's health care system. *Health Economics* 14: 101–118.
26 Stig, K., I. P. Lütz, and the Nordic Reference Group. 2013. *Financing health care in the Nordic countries*. Copenhagen. Denmark: Nordic Medico Statistical Committee, p. 53.
27 World Health Organization. 2012. National profile of occupational health system in Finland. Copenhagen, Denmark: Regional Office for Europe, p. xii.
28 Finnish Institute of Occupational Health. 2014. The status of occupational health services in Finland and the role of the Finnish Institute of Occupational Health in the development of occupational health services: Report for the international evaluation of the Finnish Institute of Occupational Health (FIOH), p. 5.
29 Finnish Institute of Occupational Health. The status of occupational health services in Finland and the role of the Finnish Institute of Occupational Health in the development of occupational health services, p. 3.
30 Teperi, J., M. E. Porter, L. Vuorenkoski, and J. F. Baron. 2009. The Finnish health care system: A value-based perspective. *SITRA Reports 82*. Helsinki, Finland: 43.
31 Vuorenkoski, L., P. Mladovsky, and E. Mossialos. 2008. Finland: Health systems review, p. 3.
32 Finnish Institute of Occupational Health. The status of occupational health services in Finland and the role of the Finnish Institute of Occupational Health in the development of occupational health services, p. 4.
33 Perkiö-Mäkelä et al. 2006. Work and health survey 2006, *Table Report*. Tampere, Finland: Finnish Institute of Occupational Health.
34 Vuorenkoski, L., P. Mladovsky, and E. Mossialos. Finland: Health systems review, p. 1.
35 Ibid. p. 4.
36 Ibid. p. 50.
37 Stig, K., I. P. Lütz, and the Nordic Reference Group. *Financing health care in the Nordic countries*, p. 55.
38 Vuorenkoski, L., P. Mladovsky, and E. Mossialos. Finland: Health systems review, pp. 43–44.
39 Jakubowski, E. and R. Busse. 1998. Health care system in the EU: A comparative study. Luxembourg: Directorate General for Research, European Parliament, p. 49.
40 Vuorenkoski, L., P. Mladovsky, and E. Mossialos. Finland: Health systems review, p. 43.
41 Ibid. p. 56.

42 Ibid. p. 51.
43 Stig, K., I. P. Lütz, and the Nordic Reference Group. *Financing health care in the Nordic countries*, p. 47.
44 Teperi, J., M. E. Porter, L. Vuorenkoski, and J. F. Baron. The Finnish health care system: A value-based perspective, p. 43.
45 Emanuel, E. and T. Spiro. 2015. The coming health-care cost shock, *The Wall Street Journal*, July 8, p. A11.
46 Doupi, P. et al. E-health strategies country brief: Finland, p. 11.
47 Ibid. p. 14.
48 Ibid.
49 Graig, L. A. Health of nations: An international perspective on U.S. health care reform, p. 180.
50 CIA World Factbook: http://www.cia.gov/library/publications/the-world-factbook/geos/fi.html. Accessed on August 8, 2015.
51 Lupton, D. 1995. *The imperative of health: Public health and the regulated body*. Thousand Oaks, CA: Sage, p. 51.
52 Ministry of Social Affairs and Health. 2011. *Challenges and opportunities of the 2010s*. Helsinki, Finland, p. 6.
53 Simnett, I. 1995. *Managing health promotion: Developing healthy organizations and communities*. New York, NY: John Wiley & Sons, p. 2.
54 Lupton, D. *The imperative of health: Public health and the regulated body*, p. 51.
55 Teperi, J., M. E. Porter, L. Vuorenkoski, and J. F. Baron. 2009. The Finnish health care system: A value-based perspective, p. 112.

7 Technopolitan Society

Enduring Value Change

This chapter focuses on the role of information technology and culture in addressing critical issues of public policy and governance. The emphasis will be on how technology can drive culture and value change thereby casting specific influences on the policy assessment process both in terms of a general civic trust of the national government and in the efficacy of specific programs, particularly in education and health policies. How does technology use mediate between these program assessments and other factors that may affect the future of society? The above issues are framed with the understanding that through technology-enhanced knowledge practices, learning by collaborative designing, Internet/Intranet, mobile/fixed communication systems or regional information systems that serve as avenues for knowledge generation and diffusion, citizens more or less become aware of what the government is doing and how its programs are able to solve most social and administrative issues of political governance. The main research question therefore revolves around how increasing information technologies (ICT) use shape or change civic attitudes in the evaluation of the effectiveness of program or policy outcomes, as well as in citizens trust of the national government.

7.1 Data and Hypotheses Development: Basic Assumptions

The information for this study is taken from 442 surveys disseminated to various agencies and individuals across selected localities in Finland. Of the total, approximately 107 surveys were completed and returned constituting about 24% response feedback. Considering the expanse of territory covered and the prohibitive logistics of covering every nook and cranny, I find this number robust enough for this study and likely to yield useful and generalizable results. Based on the initial proposition that an increasing level of technological application in Finnish life may also generate attitudinal changes and novel perspectives about critical political and social problems, three general sets of hypotheses are developed:

H1: Increasing levels of ICT utilization is more likely to influence civic attitudes in their trust of the national government and its policies.

H2: Sociopolitical differences are more likely to influence civic attitudes regarding their evaluation of educational and health policy outcomes.

H3: Sociopolitical differences are more likely to influence civic attitudes regarding trust of the national government and its policies.

7.2 Research Design and Methodology

The study utilizes both qualitative and quantitative approaches as a way of addressing the practical and normative issues implied in this study. Data collection was based on a survey/questionnaire instrument that was administered across the country of Finland in the fall of 2013. The questionnaire was duly approved by the Institutional Review Board (IRB) of Auburn University, Montgomery (Protocol No. 2013–046) and by the Academic Ethical Committee of the Tampere Region located at the University of Tampere, Finland, in conformity with the informed consent provisions and federal regulations regarding participation of human subjects in a research study. The questionnaire was prepared in English and also translated into the Finnish language (see Appendices B and C). It was disseminated randomly by mail or physically to members of the Parliament, personnel of the Ministries of Education and Culture, the Ministry of Social Affairs and Health, administrators in select local/municipal government jurisdictions and randomly to the general public in various cities and localities such as Helsinki, Turku, Tampere, Oulu, Rovaniemi, Espoo, Jyväskylä, Vaasa, Joensuu, Hämeenlinna, and others.

Second, in assessing the benefits of several visits to the Finnish Parliament, the Ministry of Education and Culture, the Ministry of Social Affairs and Health, SITRA-The Finnish Innovation Fund, the National Institute of Health and Welfare, the Ministry of Foreign Affairs, the Finnish National Board of Education, the Finnish Communications Regulatory Authority (FICORA), the Library of Parliament, Nokia Headquarters in Espoo, and the Parliament's Committee for the Future, one gets the idea of a political system in which equality of outcome (at least in expectation) transcends all aspects of public policy both in design and in implementation. It is with this in mind that one can feel equally confident that the survey instrument as designed would be able to capture some of the general and more specific nuances or differences in how the public or citizens of Finland make particular judgments about their national government, public policies or the broader issues facing mankind and the future.

At the policy outcomes level, we can expect to get a better understanding of how technology, ideology, or socioeconomic differences help shape the kind of civic values that citizens generally attach to the efficacy of political institutions, as well as in their assessment of the effectiveness of select public programs in addressing key social problems that fall under their programmatic jurisdictions. The questionnaire information was coded in such a way that it would lend itself to a factor analysis procedure as way of detecting key underlying dimensions or phenomena that cut across the various participant responses and also in what ways cultural or ideological dispositions and attitudes may converge or diverge between the responses. Factor scores can be extracted from the survey data as a way of delineating how key explanatory variables validate the hypothesized relationships in the overall model specification.

In analyzing the data, three sets of measures were used as a way of getting a general indication of how the responses spread out or come together on the various segments delineated in the questionnaire. The first test was in determining possible differences in the mean rank of the responses based on respondent's ideological preference, gender, and education by using the nonparametric measure in the form of the Mann-Whitney *U* test.[1] The test allows for a determination of whether two sets of responses come from the same distribution, and because it uses ordinal rather than interval level data, it can be considered the equivalent of an independent sample *t* test for comparing two nominal groups or categories. The second process used the cross-tabulation procedure and the Chi-square (χ^2) test of significance. Additionally, the analysis of variance (ANOVA) was employed as a way of determining the difference in the mean responses of the various categories of respondents.

7.3 Technoculture: Evaluation and Analysis

While there are many ways to try to make sense of the distribution of responses in the data, I have chosen a more conservative and parsimonious approach but one that offers crucial pathways for understanding changes and stability in the average Finnish attitude regarding issues of practical political governance as well as the challenges of the larger global community. The information in Table 7.1 shows that only the variable "What is the greatest national threat?" was significant at $p < .05$ level. What this points to is that large differences between the mean ranks of Conservatives and Liberals were indicated by a lower *U* score. Conservatives reported significantly higher concern about the existence of national threats than respondents who are liberal. This also conforms to the oft-perceived notion that Conservatives are more likely to be more emphatic about national security issues than Liberals. Nonetheless, there were no large differences in the mean ranks for the rest of the variables. The additional two-sample Kolmogorov-Smirnov test was insignificant, confirming that the mean ranks for both Conservatives and Liberals are normally distributed, thus accepting the null hypothesis of normality.

Table 7.1 A Comparison of the Mean Rankings of Respondents based on Differences in Ideological Preference (Mann-Whitney *U* test), N = 54

Variable Name	*U*	*Mean Rank Conservatives*	*Mean Rank Liberals*	*p < .05*
Trust National Government	251.0	27.58	27.48	.982
Foremost Problem in Finland	248.5	27.79	27.42	.941
Foremost Global Problem	221.5	30.04	26.77	.512
Greatest National Threat	151.0	19.08	29.90	.026*
Effectiveness Education Policy	217.0	30.42	26.67	.420
Effectiveness Health Policy	219.5	30.21	26.73	.464

None of the variables in Table 7.2 was significant and the additional two-sample Kolmogorov-Smirnov test was insignificant, thus confirming that the mean ranks for both male and female are normally distributed, thus accepting the null hypothesis of normality. The same conclusion is reached for the variable that measured the responses based on differences in educational status (Table 7.3). The additional two-sample Kolmogorov-Smirnov test was insignificant confirming that the mean ranks for High School Diploma (HSD) and Bachelor's Degree (BD) are normally distributed, thus accepting the null hypothesis of normality.

Table 7.2 A Comparison of the Mean Rankings of Respondents Based on Differences in Gender (Mann-Whitney *U* test), N = 107

Variable Name	*U*	*Mean Rank Male*	*Mean Rank Female*	*p < .05*
Trust National Government	1361.5	55.27	52.89	.671
Foremost Problem in Finland	1320.0	56.10	52.16	.505
Foremost Global Problem	1364.5	52.79	55.06	.697
Greatest National Threat	1294.5	56.61	51.71	.391
Effectiveness Education Policy	1177	49.04	58.35	.091
Effectiveness Health Policy	1308	56.34	51.95	.436

Table 7.3 A Comparison of the Mean Rankings of Respondents Based on Differences in Education (Mann-Whitney *U* test), N = 61

Variable Name	*U*	*Mean Rank HSD*	*Mean Rank BD*	*p < .05*
Trust National Government	352.0	32.70	27.76	.267
Foremost Problem in Finland	298.0	34.05	25.19	.060
Foremost Global Problem	391.5	31.71	29.64	.655
Greatest National Threat	357	32.58	28.0	.317
Effectiveness Education Policy	365	32.38	28.38	.355
Effectiveness Health Policy	417	31.08	30.86	.959

HSD = High School Diploma   BD = Bachelor's Degree

7.4 Value Change: A Research Note

In terms of "Trust of the national government," women tend to be more trusting of the national government than men in the range of 57% (most of the time) compared to 32% for males (Table 7. 4). The χ^2 was significant at p = .00 <.05, and the Cramer's V of .001 suggests complete independence between male and female responses on this particular variable. "The chi-square (χ^2) test of significance determines whether the observed dispersal of cases (responses) departs significantly from what we would expect to find if the null hypothesis were correct."[2] In this case, the null hypothesis suggests that the population (males and females) does not differ on its relative trust of the national government and that the true mean difference is 0. Because the p-value is equal to or less than .05, the null hypothesis is thus rejected. Based on the respondents, women in Finland are more likely to be more trusting of their national government than males. Figure 7.1 further illustrates the difference between the genders.

On the question of whether differences in ideological preference had a relationship to respondents' trust of the national government, about 58.3% of Conservatives indicated "most of the time," while 47.6 % of Liberals indicated the same, the Socialists 52% and Egalitarians 25%. The chi-square test (χ^2 = .157) was not significant at the (p > .05) level, hence accepting the null hypothesis that there was no difference in the distribution of the mean responses of the respondents when it comes to the issue of trusting the national government, hence ideological differences play little or no role. As there were more Liberals in the sample of respondents, one would have expected a greater disposition toward the national government and its social programs. Nonetheless, this could possibly be attributed to the egalitarian and equalitarian nature of Finnish society as well as citizens' attitude toward the welfare state system (Figure 7.2).

In terms of differences in educational attainment (Figure 7.3), about 40% of respondents with high school diploma indicated that they trust the national government "most of the time," while 32% of them stated that they trust that national government "some of the time." About 57.1% of respondents who had Bachelor's degrees indicated that they "trust the national government," while 19% indicated that they did so "some of the time." About 50% of those with Master's degrees indicated "most of the time" while 29% indicated "some of

Table 7.4 Opinions on Trust of the National Government and Its Policies, by Gender

Trust the National Government	*Male*	*Female*	*Total*
No Answer	2 (4.0%)	0 (0.0%)	2 (1.9%)
Just about Always	2 (4.0%)	8 (14%)	10 (9.3%)
Most of the Time	16 (32%)	33 (57.9%)	49 (45.8%)
Some of the Time	19 (38%)	13 (22.8%)	32 (29.9%)
Hardly Ever	10 (20%)	1 (1.8%)	11(10.3%)
Don't Know	1 (2.0%)	2 (3.5%)	3 (2.8%)
Total Count	50	57	N = 107
χ^2	p = .001	Cramer's V = .001	λ = .612

Note: Percentages are in parenthesis.

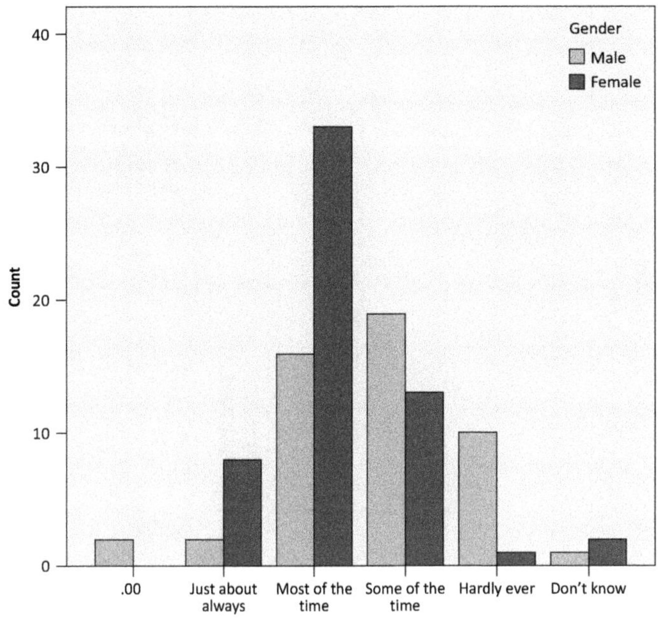

Figure 7.1 Trust of the National Government, Determined by Differences in Gender.

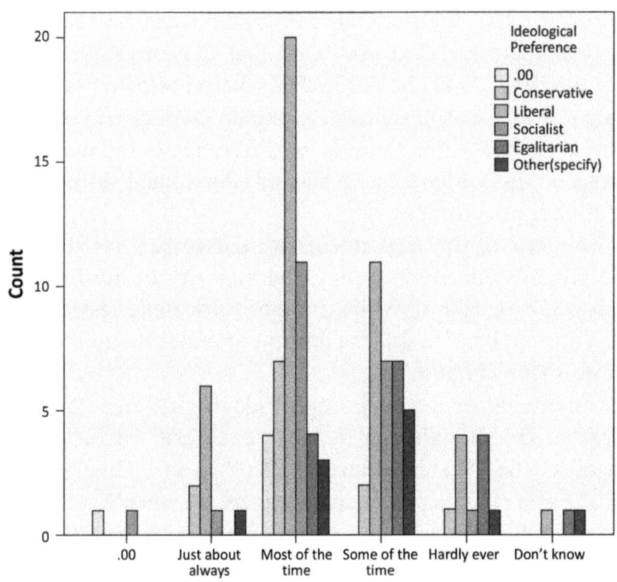

Figure 7.2 Trust of the National Government, by Ideological Preference.

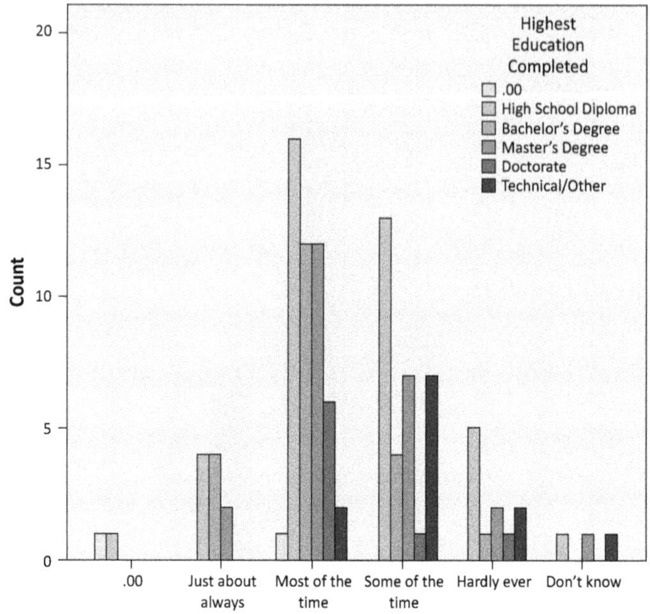

Figure 7.3 Trust of the National Government, Determined by Differences in Educational Status.

the time." But the most noticeable impact on the role of education in informing judgments about trusting the national government can be seen in that 75% of respondents with doctorate degrees stated that they trust the national government "most of the time," and about 12.5% of them indicated "some of the time." The chi-square test (χ^2) was significant at the.05 level ($p = .00 < .05$). This thus rejects the null hypotheses that there is no difference in the mean distribution of responses based on the various categories of educational status as they relate to trusting the national government.

When the opinions of the respondents were assessed based on how effective they viewed Finland's education policy, the majority of the female respondents thought that it was "excellent" while a majority of male respondents thought it was "average," although a sizeable proportion of males thought that the education policy was "excellent" (Figure 7.4).

When measured against the backdrop of ideological differences (Figure 7.5), 50% of Conservatives thought that it was "excellent," as opposed to Liberals (52.4%), Socialists (66.7%), Egalitarians (25.0%), and Others (18.2%). With a chi-square of .00 at $p < .05$ level of significance, we must reject the null hypothesis that there is no difference in the mean responses on the effectiveness of educational policy measured against existing ideological differences.

In terms of the effects of educational differences in how respondents evaluated the effectiveness of educational policy (Figure 7.6), the responses indicate that 60% of those with high school diplomas rate the educational policy as "excellent," Bachelor's degrees (38.1%), Master's degrees (37.5%), Doctorate

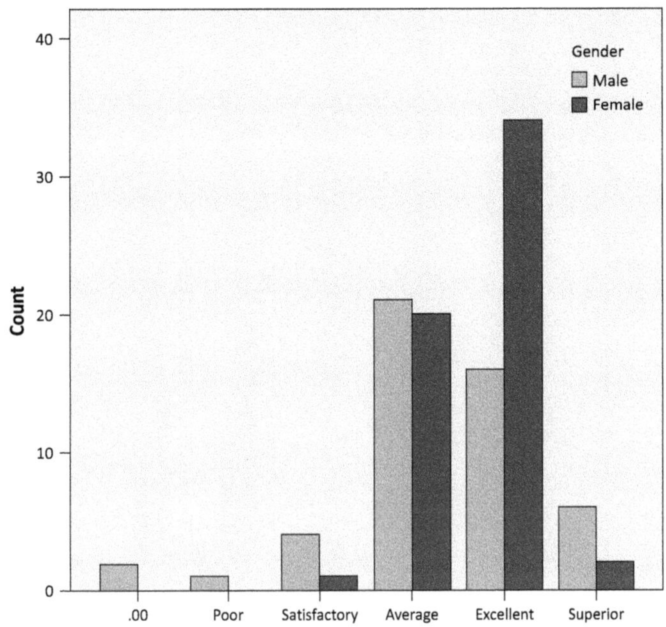

Figure 7.4 Effectiveness of Education Policy, by Gender.

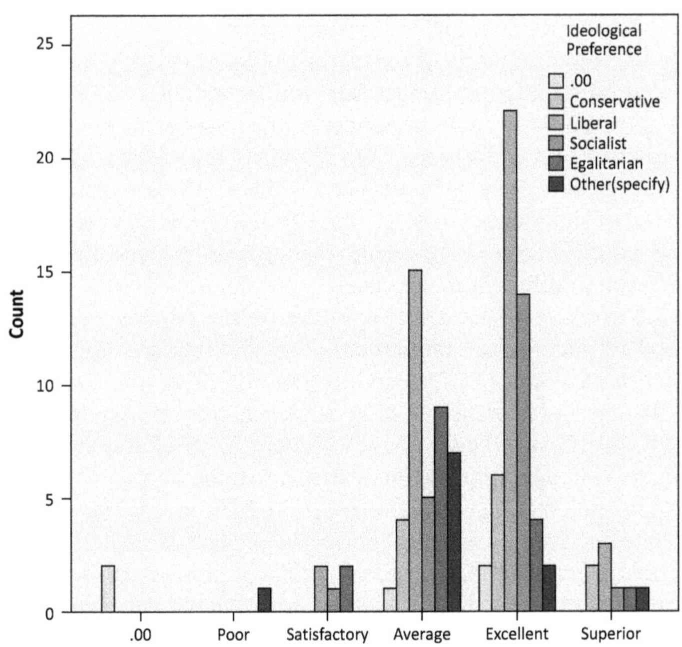

Figure 7.5 Effectiveness of Education Policy, by Ideological Preference.

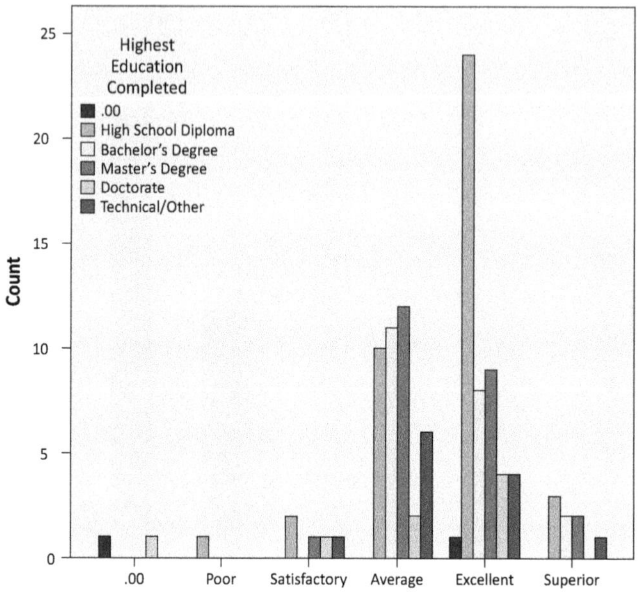

Figure 7.6 Effectiveness of Education Policy, by Differences in Educational Status.

degrees (50%), and Technical/Vocational diplomas (33.3%). With a chi-square at ($p = .00 < .05$), we must reject the null hypothesis that there is no difference in the mean responses regarding the effectiveness of education policy when considered in light of differences in educational status.

But when assessed in terms of the effectiveness of health policy (Figure 7.7), there were no noticeable differences between the opinions of males and those of females. While 32% of male respondents rated the health policy as excellent, about 40% also rated it as "average." On the other hand, 29.8% of female respondents rated the health policy as "excellent," while 47.4% saw it as "average." The chi-square test of significance was .215 ($p > .05$), so we accept the null hypothesis that there is no difference in the mean responses on the effectiveness of health policy as a result of differences in gender.

Figure 7.8 assesses respondents' opinions on the effectiveness of the health policy based on ideological differences. The distribution shows that Conservatives rated it excellent (41.7%) and average (25.0%) while Liberals rated it excellent (28.6%) and average (54.8%); Socialists rated it excellent (47.6%) and average (33.3%); Egalitarians rated it excellent (25.0%) and average (37.5%). The chi-square test of significance indicated a p-value of .09 ($p > .05$), hence we must accept the null hypothesis that there is no difference in the mean responses regarding the effectiveness of health policy when considered in light of ideological differences. Equally, differences in political or partisan affiliation ($\chi^2 = .287$; $P > .05$) was of no influence in the mean responses in the assessment of the effectiveness of health policy. It did not matter as much whether a respondent was

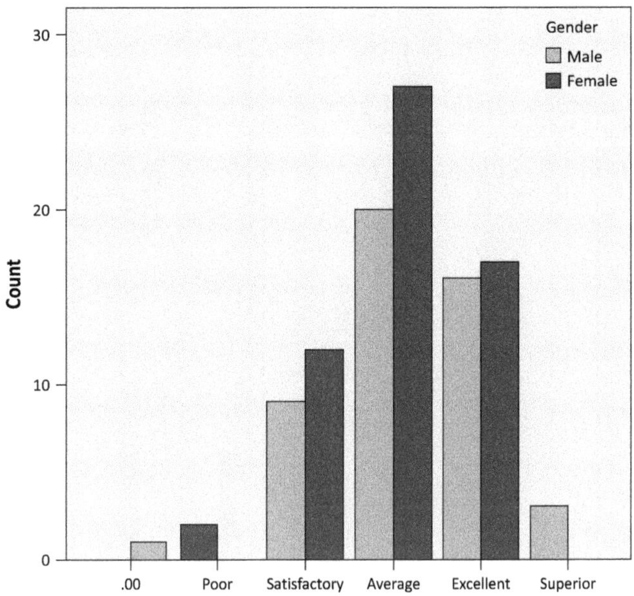

Figure 7.7 Effectiveness of Health Policy, by Differences in Gender.

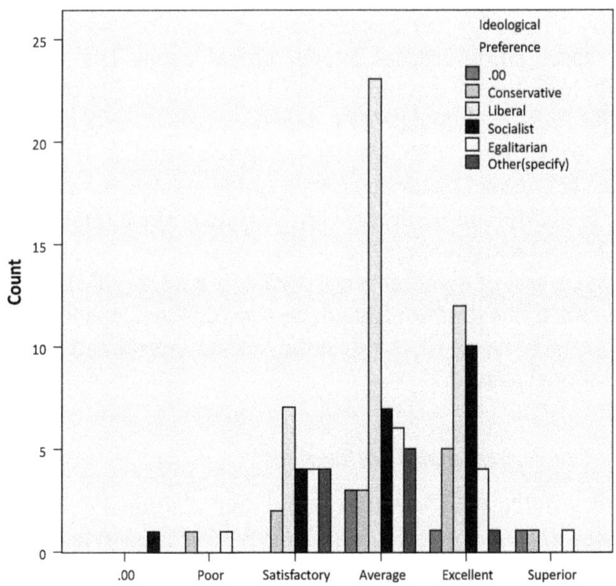

Figure 7.8 Effectiveness of Health Policy, by Ideological Preference.

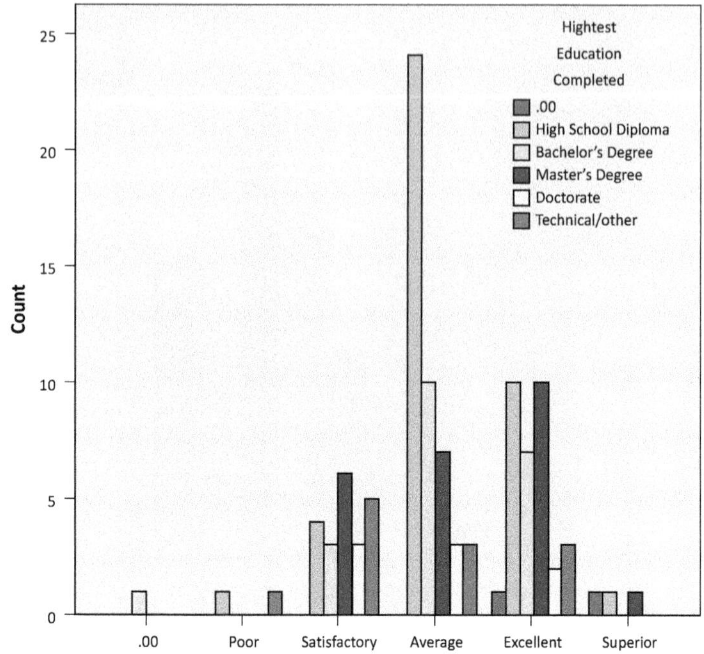

Figure 7.9 Effectiveness of Health Policy, by Differences in Educational Status.

a member of the Center Party, Green League, Left Alliance, National Coalition Party, Social Democratic Party, Swedish Peoples Party or Independent.

Additionally, Figure 7.9 assesses respondents' opinions on the effectiveness of health policy based on differences in educational status. The distribution shows that those with high school diplomas rated it "excellent" (25.0%) and "average" (60.0%), while those with Bachelor's degrees rated it "excellent" (33.3%) and "average" (47.6%); those with Master's degrees rated it "excellent" (41.7%) and "average" (29.2%); Doctorate degree holders rated it "excellent" (25.0%) and "average" (37.5%), while those with Technical/Vocational diplomas rated it "excellent" (25.0%) and "average" (25.0%).

The chi-square test of significance indicates a p-value of .03 ($p < .05$), hence we must reject the null hypothesis that there is no difference in the mean responses regarding the effectiveness of health policy when considered in light of differences in educational status.

7.4.1 Testing Convergence and Differences

As a means of providing further elaboration and additional confirmatory assessment, a one-way analysis of variance (ANOVA) was used to test for the homogeneity of variance (Levene statistic) as well as significant differences in the mean responses. The process involved the use of three independent variables (gender, educational status and ideological preference) and a group of dependent variables (items) as indicated in Table 7.5. (The between and within group sum of squares and the F-test have been deliberately omitted.)

Based on the analysis of variance, the variables of note were respondents' trust of the national government, assessment on the effectiveness of the education policy and opinions on the foremost global problem facing the world. Because the mean responses on respondents' "trust of the national government" based on all categories of the independent variables (gender, educational status, and ideological preferences) was significant at the $p < .05$ level, we can thus reject the null hypothesis that there is no difference in the mean responses regarding respondents' trust of the national government when considered in light of differences in ideological preferences. On this score, the respondents differ significantly in their degree of trust of the national government. However, the puzzling difference between this result and the result from the earlier chi-square test of significance could be due to the fact that the distribution of the respondents by ideological preference alone seemed to weigh heavily in favor of those who indicate themselves as Liberals (39.3%) as opposed to Conservatives (11.2%), Socialists (19.6%), Egalitarians (15%) and others (10.3%). It could also be due, perhaps, to the fact

Table 7.5 ANOVA: Mean Differences on Likert-Scale Dependent Variables, By Gender, Education and Ideological Preference

	Dependent Variables/Items	*Gender*	*Independent Variables Education*	*Ideology*
1.	Trust of the National Government	6.44 (.00)*	2.64 (.012)*	2.83 (.00)*
2.	Frequency of Internet Usage	3.86 (.121)	.592 (.876)	2.54 (.157)
3.	Media Influence on Government	.135 (.797)	4.14 (.065)	3.97 (.077)
4.	Effectiveness of Education Policy	2.88 (.05)*	1.51 (.079)	2.69 (.00)*
5.	Effectiveness of Health Policy	.559 (.397)	1.45 (.089)	1.40 (.103)
6.	Greatest National Threat	2.08 (.411)	1.45 (.803)	2.47 (.552)
7.	Foremost Global Problem	.535 (.687)	5.83 (.106)	7.56 (.036)*
8.	Foremost Problem in Finland	2.21 (.419)	3.09 (.471)	.725 (.958)
9.	Need to Preserve Cultural Norms	.135 (.746)	2.31 (.102)	1.67 (.255)
10.	Technology Complicates Understanding	.317 (.586)	.803 (.585)	1.64 (.165)
11.	Technology more Good than Bad	1.68 (.219)	1.83 (.137)	1.09 (.425)

Whole numbers = Mean Squares Parenthesis = p-value (.05) Asterisk = Sig. at $p < .05$ N = 107.

that the analysis of variance procedure is a relatively more statistically sensitive approach than the chi-square.

The data also show that females tend to be more trusting of the national government and its policies than males (Figure 7.1), possibly as a result of the egalitarian character of Finnish society and its emphasis on universal equality. In terms of differences in educational status, those with high school diplomas tend to be more trusting of the national government than those with Bachelor's or Master's degrees (Figure 7.3). In a way, this can be a reflection of the universal education system and the central emphasis given to basic education as well as vocational training at the lower educational levels. For differences in ideological preference, the data show that Liberals tend to be more trusting of the national government than Conservatives or Socialists (Figure 7.2).

In terms of how the respondents assessed the effectiveness of educational policy, the mean responses based on gender and ideological preferences were significant at $p < .05$, hence we must reject the null hypothesis that there is no difference. Based on Figure 7.4, more females than males thought that education policy in Finland was "excellent," and in Figure 7.5, more Liberals than Conservatives and Socialists viewed the education policy as "effective." Hence, differences in gender and ideological preferences can explain the different ways in which the respondents evaluated the effectiveness of the country's education policy as well as individual dispositions or attitude toward what they saw as the foremost problem(s) facing the world. For the rest of the variables in Table 7.5 that were not significant at $p < .05$, we can thus accept the null hypothesis that there is no difference in the mean responses and any observed variation can most possibly be due to chance. But overall, the results lend greater support to the three hypothesized relationships and the various propositions they represent in this study.

On the question that asked respondents to indicate what they thought to be the "foremost global problem," more Liberals than the other ideological preferences indicated "global hunger, climate change, and war and conflict," in that order (Figure 7.10). It is also noteworthy to point out that such issues as drug and human trafficking, transborder immigration, and even globalization ranked very low in terms of what respondents thought were major problems facing the world.

As a further illustration, a corollary question was asked about what were the "foremost problems in Finland." Figure 7.11 indicates that based on gender differences, the foremost problems in Finland are globalization, climate change and culture change. Technology, transborder immigration, and drug and human trafficking ranked at the lowest end.

Considered in terms of differences in educational status (Figure 7.12), the same factors (globalization, climate change and culture change) were indicated as the foremost problems in Finland, and this same result cuts across all educational levels (high school diploma and Bachelor's and Master's degree holders).

But when respondents were assessed in terms of the frequency of Internet use (Figure 7.13) and its communicative efficacy, the results show that frequency of Internet use was associated with trust of the national government. About 51% of the respondents indicated that they use the Internet "always," while 22% use it

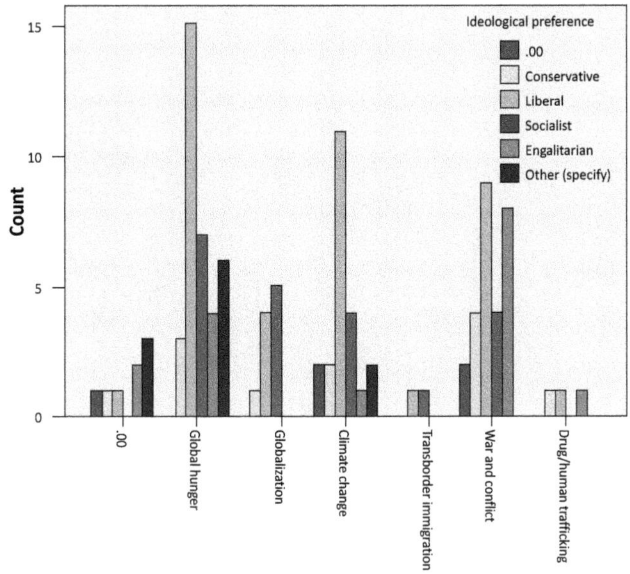

Figure 7.10 Foremost Global Problem, by Ideological Preference.

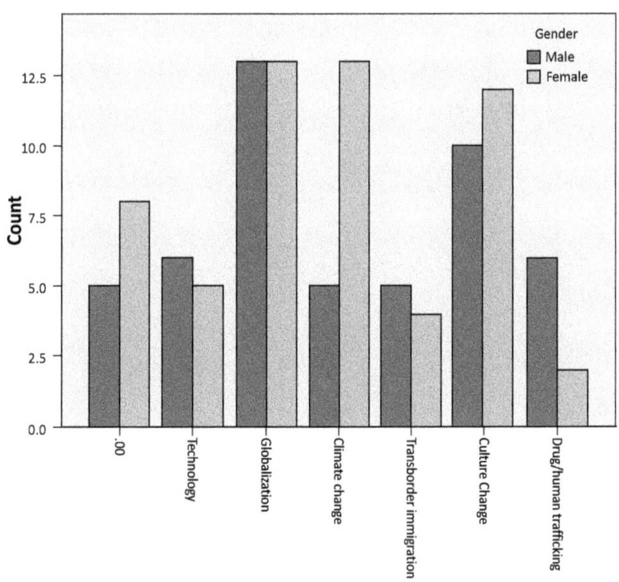

Figure 7.11 Foremost Problem in Finland, Determined by Gender.

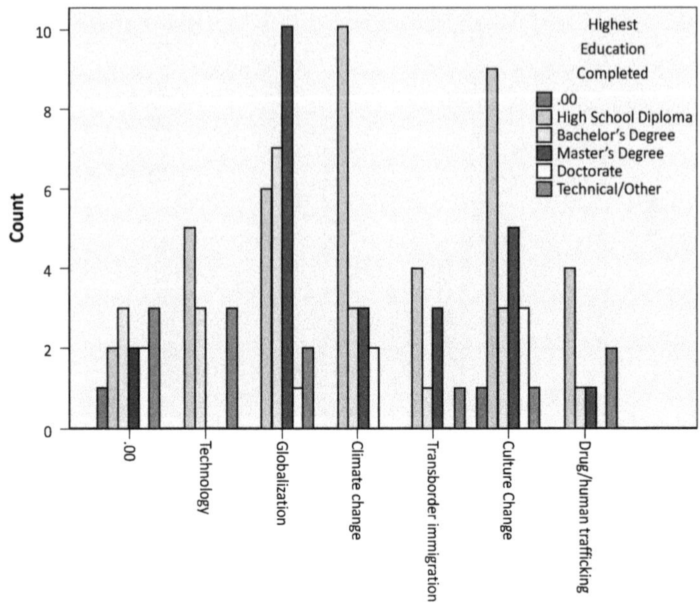

Figure 7.12 Foremost Problem in Finland, Determined by Differences in Educational Status.

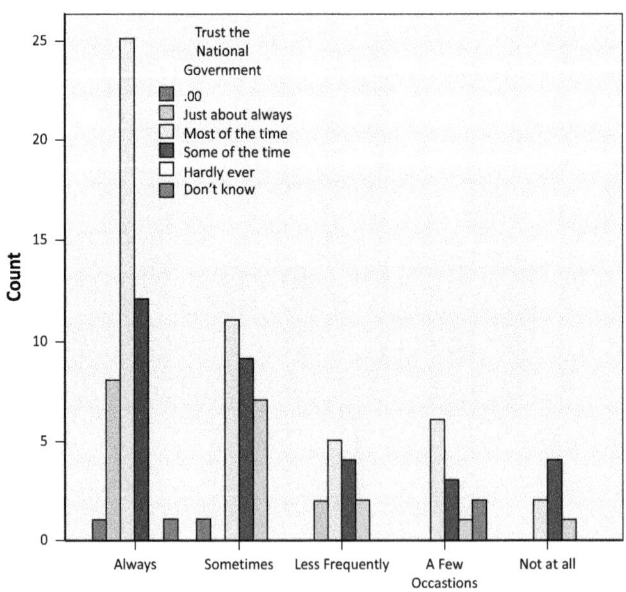

Figure 7.13 Trust of the National Government, Determined by Frequency of Internet Usage.

"sometimes," and 10.2% use it "less frequently." The chi-square (χ^2) value of .03 ($p < .05$) was significant, hence we can reject the null hypothesis that there is no difference in the mean responses for trust of the national government when considered in light of the frequency of Internet use. People who always use the Internet or use it "sometimes" are more trusting of the national government than those who use the Internet less frequently. And the most used medium of communication for them was e-mail and cellphones.

It could thus be said that since much of the interaction between citizens and government tended to occur over the Internet and e-mails, the lack of continuous face-to-face contacts may have de-sensitized much of the public into casting a more personalized judgment about the government and the effectiveness public officials bring to the execution of their duties. While it may seem that "the Internet and e-mail function mainly to enhance the administrative relations and easy-to-use interfaces between citizen and government, the unequalled penetration of mobile phones among citizens worldwide may only have demonstrated the speed of technological diffusion."[3] Nonetheless, the above finding presents a rather interesting puzzle in terms of ICT utilization, particularly the Internet in communications with specific public agencies (Figures 7.14 and 7.15).

Considered in terms of differences in gender, both male and female respondents overwhelmingly indicated that they have not accessed or communicated with the Ministry of Education and Culture and the Ministry of Social Affairs and Health over the Internet (including of course accessing critical public information on their websites).

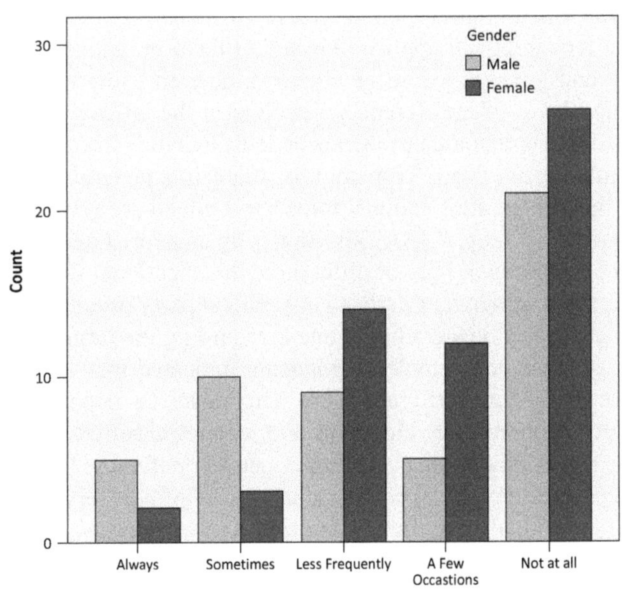

Figure 7.14 Internet Access to the Ministry of Education and Culture, Determined by Gender.

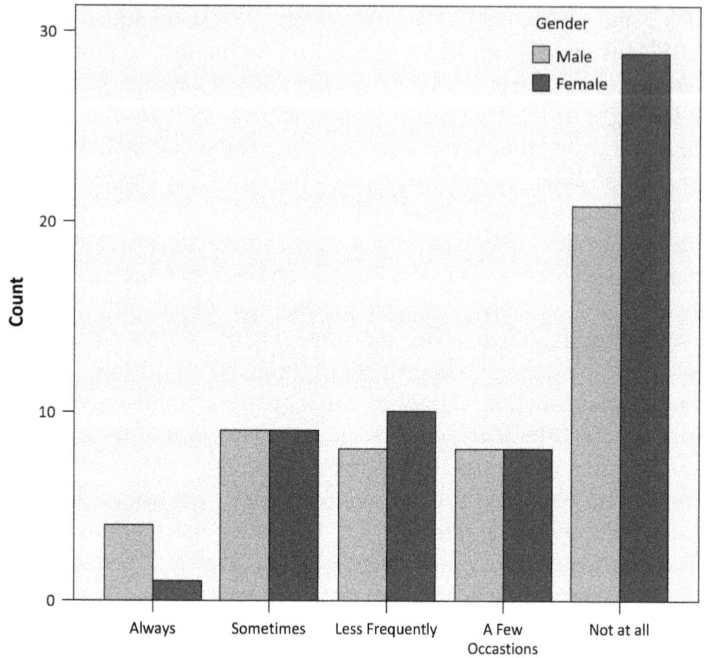

Figure 7.15 Internet Access to the Ministry of Social Affairs and Health, Determined by Gender.

To the extent that this study characterizes Finland as a *technopolitan society* (in the sense that technological applications and utilization ubiquitously span every aspect of life and governance in the country), does this low level of utilization suggest a rising level of indifference, a decline in the efficacy of technology in solving most social and human problems or self-alienation due to technology burden or information overload? Nonetheless, there is a general consensus across gender lines that the greatest national threats to Finland are economic depression and unemployment (Figure 7.16). This view is by an overwhelming majority also upheld when considered in light of differences in educational status and ideological preferences and in terms of partisan or political party orientation.

Also on the issue of how confident one is regarding the future, an overwhelming majority of male and female respondents indicated that they have little or no confidence in the future (Figure 7.17). This raises an important question. To what extent are economic development and technological advancement related to increasing levels of optimism, happiness and hopefulness? Future economic development and technological advancement are helpful in serving the needs and well-being of individuals and society, but they also have costs.

What can thus be gleaned from these analyses is that while Finns generally share similar egalitarian and social welfarist values, they do not walk in lock-step. It also suggests that the ever-present role of the liberated and free individual,

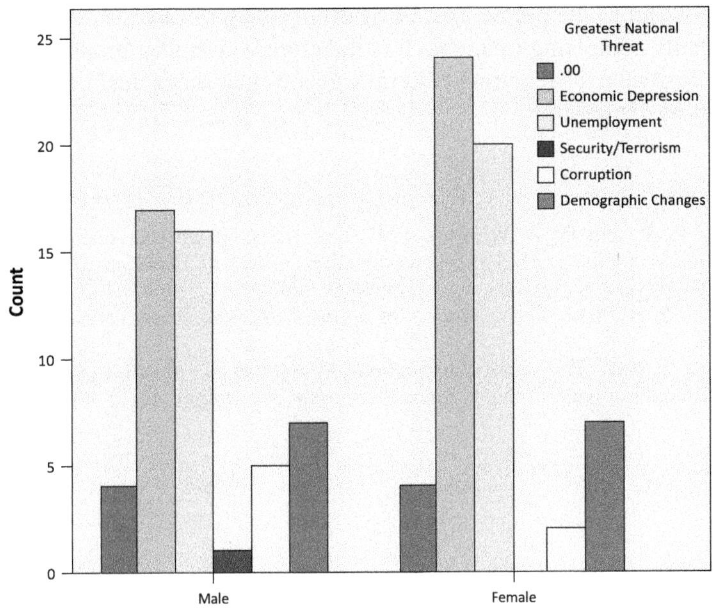

Figure 7.16 Greatest National Threat, Determined by Gender.

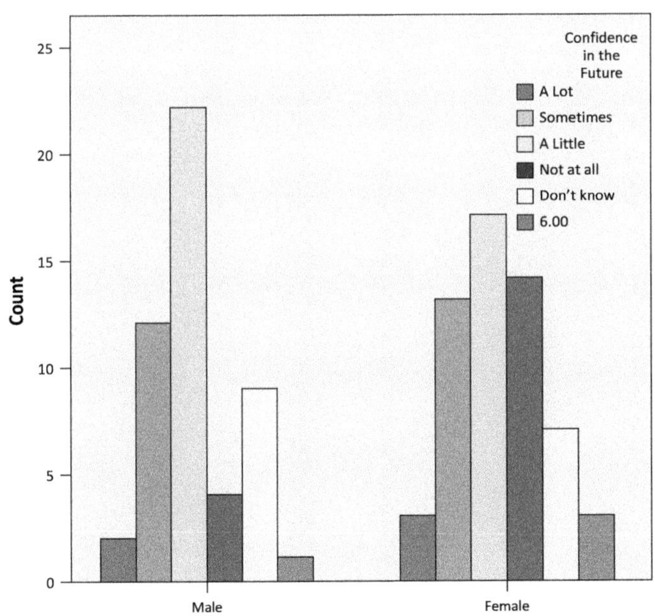

Figure 7.17 Confidence in the Future, Determined by Gender.

though mindful of the public interest, is ever-willing to cast his own views amid other equally competing opinions. It is therefore within this template that one is able to experience the enduring cultural vignettes that shape the Finnish character as well as its expression.

Notes

1 Mann, H. B. and D. R. Whitney. 1947. One a test of whether one of two random variables is stochastically larger than the other. *Annals of Mathematical Statistics,* 18: 50–60.
2 Pollock, P. H. 2012. *The essentials of political analysis*. New York, NY: CQ Press, p. 164.
3 Inkinen, T. 2006. The social construction of the urban use of information technology: The case of Tampere, Finland. *Journal of Urban Technology*, 13(3): 49–75, p. 66.

8 Evolving Social Structure

Resiliency and Change

It is a known fact that any process of development involves change, in the way things are generally done or in the habits, norms, beliefs and reflections people bring to bear on mundane decisions and everyday activities. There are objective and subjective aspects of change, the objective being more visible and the subjective more internalized and possibly invisible at least in the more overt sense of the term. But from a sociological perspective, it can be argued that the subjective aspect of change can be more enduring and even potent in shaping the outward expressions that are manifested more visibly. On the basis of epistemological considerations alone, we can thus arrive at the conclusion that "our knowledge of social phenomena is to be found in the relational connection between the observable features of social action and those unobservable structures that shape events";[1] hence, in this regard, causality can be construed as the product of the relationality of unobservable structure and the observable human agency at any given time and place.

Because cultural dispositions can be manifested at both the subjective (micro) and objective (macro) levels, they oftentimes generate reciprocal effects, hence making any direct attribution of causality, directionality and intentionality much more difficult to ascertain. There are several assumptions that can also be made, among them whether increasing levels of ICT utilization lead to a change in traditional cultures and habits, whether extant cultural dispositions dictate how and for what use technology is applied or whether such a relationship is, nonetheless, reciprocal.

8.1 Values and Practices: Defining Characteristics

It may be true that Finland is Scandinavian by geography, but it is culturally Nordic. It is not Russian or Slavic, but it is one of those rare and unique coincidences of nature that the Finns see themselves as a special breed, a culturally distinct group different from their immediate neighbors. Perhaps this will explain one of the reasons that, although Finns may appreciate other cultures (an ideal that is reflected in the grouping of independent responses loaded onto Factor 3 of Table 8.3: "Other people's ethnic values are just as good as mine"), they are still reluctant to accept these 'others' among them, as an essential element of Finnish ethno-cultural identity. They may see multiculturalism or globalization as worthy of experimenting with but not something they really need to wholly embrace, in light of the general and equally legitimate concern about problems of culture change and the need to preserve the traditional cultural norms that have made Finland unique.

This very crucial point is numerically well reflected in the aggregate responses (shown in Figures 8.1, 8.2 and 8.3). Measured across differences in gender, educational status and ideological preference, the majority of the respondents indicate that globalization and concern for culture change were among the foremost problems in Finland. About 26% and 22.8% of males and females, respectively, pointed to globalization as the most pressing problem, while 20% of males and 21% of females also pointed to culture change as the foremost problem. By virtue of the chi-square significance at ($\chi^2 = .40$, greater than $p < .05$), we can thus accept the null hypothesis that there is no difference in the mean responses between males and females on the issue of the foremost problem facing Finland. There was equally no difference in the mean responses on this issue based on differences in ideological preferences ($\chi^2 = .923$) or on differences in educational status ($\chi^2 = .336$).

Additionally, respondents' opinions were measured against the question: "*Do you believe that the traditional culture and norms of Finland should be preserved*

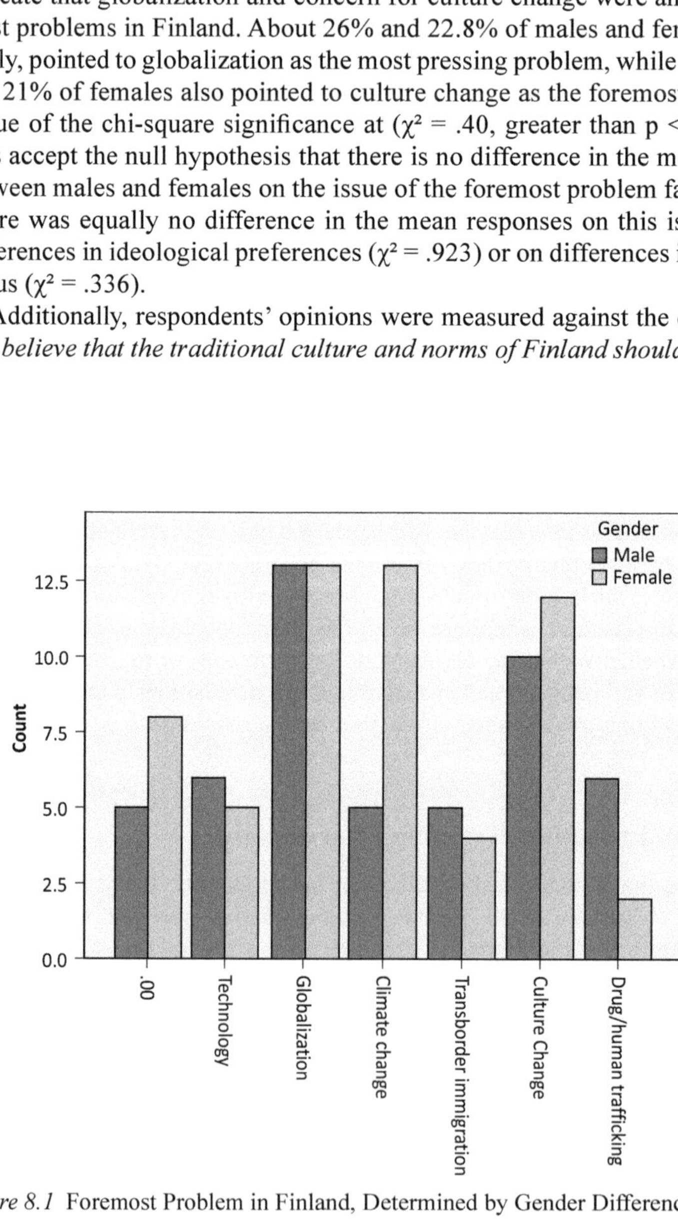

Figure 8.1 Foremost Problem in Finland, Determined by Gender Differences.

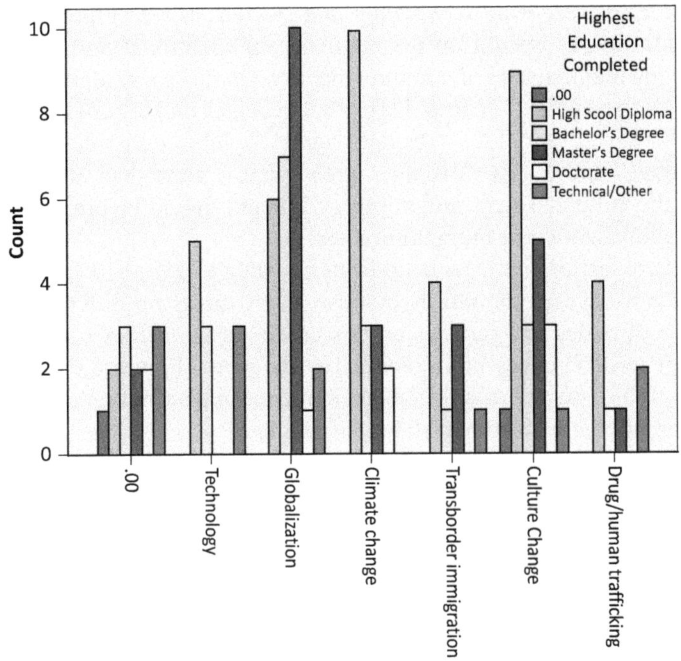

Figure 8.2 Foremost Problem in Finland, Determined by Differences in Educational Status.

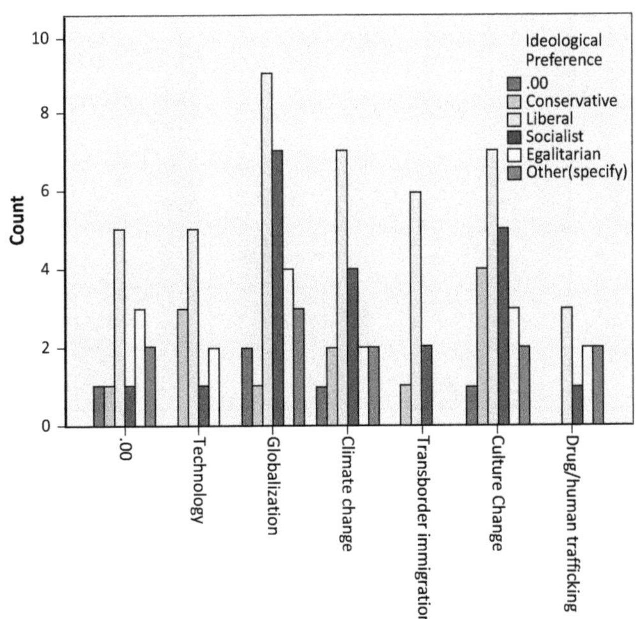

Figure 8.3 Foremost Problem in Finland, Determined by Differences in Ideological Preferences.

even in the face of increasing technology and globalization?" Figures 8.4, 8.5 and 8.6 provide further insight on the resiliency of traditional norms and cultural habits in the maintenance of common identity. On this score alone and when measured against gender differences, Figure 8.4 becomes most illustrative. About 46% of males "agree" and 12% "strongly agree"; while approximately 40.4% of females "agree" and 17.5% "strongly agree" (that is more than 50% for both genders) that the cultural values and norms of Finland should be preserved even in the face of globalization or multiculturalism.

The chi-square test of significance indicates a value of $\chi^2 = .918$ (greater than $p < .05$), hence we accept the null hypothesis that there is no difference in the mean response of males and females on this crucial issue. While a similar conclusion applies to differences in ideological preferences (Figure 8.6: $\chi^2 = .298$, greater than $p < .05$); there was a difference in the mean distribution of responses based on differences in educational status (Figure 8. 5).

For most of those who "agreed" that Finnish cultural values should be preserved, most of them had high school diplomas (47.5%) or Bachelor's degrees (76.2%), but it would seem that the higher educational status one attains (the Master's or Doctorate degrees) the more a certain level of ambivalence and/or indecision sets in. About 29.2% of Master's degree holders and 25% of Doctorates would "neither agree nor disagree" on the need to preserve the cultural norms of Finland in light of the aforementioned challenges. The chi-square test of significance shows $\chi^2 = .00$ (less than $p < .05$); hence, we reject the null hypothesis of no difference in the mean responses based on differences in educational status.

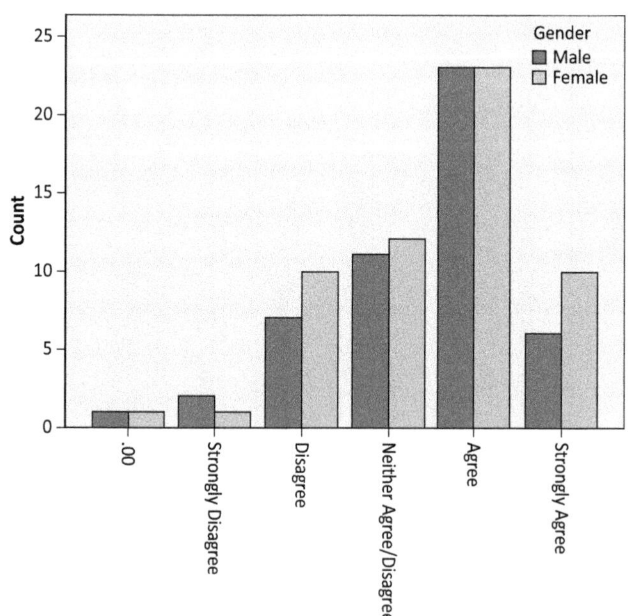

Figure 8.4 Need to Preserve Cultural Norms, Determined by Differences in Gender.

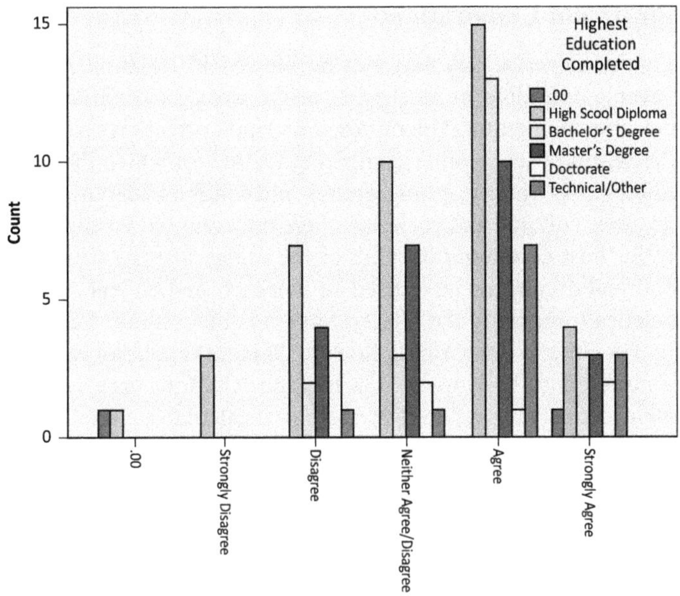

Figure 8.5 Need to Preserve Cultural Norms, Determined by Differences in Educational Status.

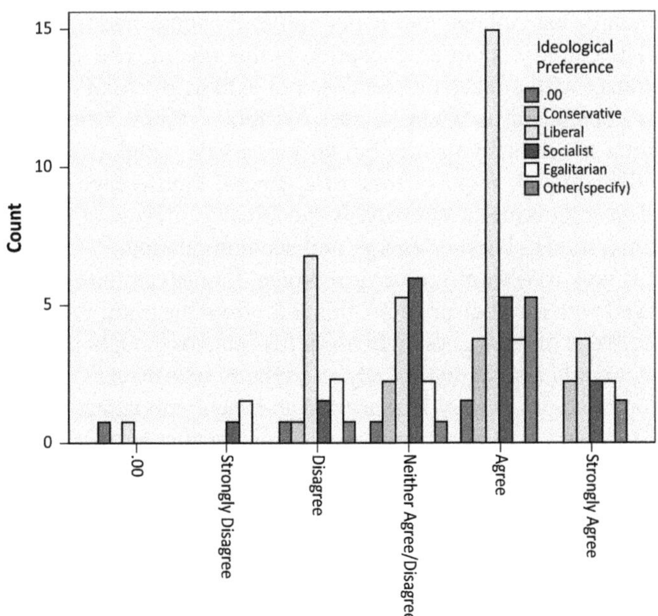

Figure 8.6 Need to Preserve Cultural Norms, Determined by Differences in Ideological Preferences.

8.2 Emphasis and Elaboration

While the above observations may not be viewed in isolation from the cross-current of events unfolding in most parts of the world, and while societies have fought to preserve their cultural identities and values, oftentimes such effort seems to work at cross-purposes to other competing factors for which there is little control or prediction. As various ethnic groups and religious identities seek accommodation in other cultures and societies, there are bound to be enduring changes, particularly for host communities. One of the things that the Chinese and other Eastern or Oriental cultures have used to entrench and endear themselves into western societies (especially the United States) is their ethnic food and eateries. While many US citizens unsparingly flock to Chinese restaurants and buffets, that allure does not seem to have worked in Finland. The Finns are very much given to their native delicacies prepared in various incalculable varieties with potato (carbohydrate), fish or meat (beef, reindeer, lamb, goose, etc.) as the central actors.

While immigrants from the Middle-Eastern and Muslim-dominant countries such as Turkey and Syria, as well as the ethnic Kurds own most of the kebab and halal eateries in Finland, they have also been creative about it by adding the more traditional pizza and pasta to their menu. In this way, they are able to attract most Finns who are more attuned to pizza and pasta than kebab or halal food. Few Africans or African-Americans live in Finland, and this can be explained by a series of factors including that Finland does not have a history of colonialism and is also geographically isolated from most of central Europe. As a political system that has made an enormous contribution to the world in various areas, Finland still seems somewhat isolated and is not culturally connected to most regions of the world. The country is not an exception in this regard. The average Finn who comes in contact with a black person does not exhibit any kind of overt curiosity but remains nonchalant as when he sees his fellow Finns. One quick eye contact invites not necessarily a smile but an immediate withdrawal to the self and an attitude that smacks, more or less, of a general indifference to the unfolding encounter. But if an issue or question is specifically directed at the Finn, he will come alive in a sudden burst of energy and accommodation.

While it is very easy to misconstrue a typical Finnish attitude as indifferent or, perhaps, tinged with racist disposition, this is far from the truth. While this does not suggest knowledge of what goes on in his or her mind or thought process (although this sentiment could also apply equally to anybody whosoever), the average Finn generally comes across as very kindhearted and accommodating, but they are also very curious to learn about each and every circumstance in which they happen to find themselves. This is what adds to the heightened curiosity whenever a visual contact is made or conversation broached between a Finn and a foreigner.

Social Capital: Shared Ideals

Social capital is based on the framework that shared values, ideas, norms and trust will create the foundation upon which collective understanding and sense of community can be realized. The concept of social capital was originally introduced by James Coleman,[2] who argued that social ties, shared norms and trust make a significant contribution to the economic efficiency and well-being of individual citizens. Over the years, however, the concept has received its most celebrated association in the works of Robert Putnam.[3] While social capital draws upon the importance of shared values and trust as critical elements in building political consensus, civic engagement and democratic discourse, Putnam[4] makes an important illustration between two forms of social capital. While he considers *bonding social capital* as inward looking and tends to reinforce existing identities among homogenous groups, he sees *bridging social capital* as outward looking and a means of creating crossover networks that transcend social cleavages. These should not be seen as opposites but rather as two dimensions that can be found in a typical social network where shared values and collective engagements have endured over time.

Table 8.1 shows that various items were strongly loaded on the specific dimension of social capital, and they collectively explain 16.38% of the variation in the underlying phenomenon. Further analysis also indicates that social capital was highly correlated with the propensity to trust the national government and its policies. Hence, the sense of increasingly shared values among the Finns may be one of the reasons a sizeable majority seems to trust the government more, and its egalitarian welfare state and emphasis on equality for all may also be reasons more females than males seem to trust the national government.

ICT Utilization

Information, communications and technology (ICT) utilization play a central role in the political, administrative and individual needs of states and citizens. ICTs can be viewed as part of a complex communication architecture that enable people and organizations to interact with each other either within or beyond their local environments. As a critical part of the ICT infrastructure, the Internet has become a tool for information exchange, maintaining social relations and increasing face-to-face interaction, all of which help to build both bonding and bridging

Table 8.1 Loadings on First Rotated Factor: Social Capital

Variables	*Factor*	*Communality*
Trust People in General	.632	.401
Trust the National Government	.638	.427
Trust Government Agencies	.678	.499
Comfortable with Co-Workers	.616	.470
Comfortable with Fellow Citizens	.667	.472
Community Comfortable with Agency Services	.730	.535
Clients/Citizens Comfortable with Agency Services	.743	.554
Shared Values with Community Members/Citizens	.538	.292
Shared Values between community and my agency	.587	.353
Confidence in the Future	.510	.275
Eigen Value	5.899	
Proportion of Variance	16.38%	
Cronbach's Alpha α = .839		

Table 8.2 Loadings on Second Rotated Factor: ICT Utilization

Variables	*Factor*	*Communality*
Frequency of Internet usage	.526	.423
Internet Access to Ministry of Education/Culture	.776	.620
Internet Access to Ministry of Social Affairs/Health	.827	.700
Evaluating Education Policy via the Internet	.759	.592
Evaluating Health Policy via the Internet	.640	.425
Eigen Value	3.57	
Proportion of Variance	9.93%	
Cronbach's Alpha $\alpha = .832$		

social capital in societies or communities.[5] Because it is assumed that stronger ties are oftentimes driven by broader media usage and ICT initiatives, the Internet provides a great opportunity for the building of interagency networks as well as the reinforcement of existing network of social relations that could eventually be accessed and mobilized for purposive action.[6] Table 8.2 show the clustering effect of several items that loaded on the specific factor or dimension of ICT utilization. Of more enduring import is that it shows ICT utilization as an essential tool in the exchange of information but also as a central element (or even a catalyst) in the management of development as well as in addressing critical issues of civic engagement and substantive policy initiatives.

Traditionalism-Identity

The traditionalist disposition also resonates aspects of a globalist cosmopolitan culture among the Finns, particularly when it comes to common interests of the international community such as arbitrating peace negotiations between parties in conflict, postwar reconstruction initiatives and peace-keeping obligations or the broader regional welfare between Finland and its neighbors, particularly Russia. For example, following the end of the Cold War, the Finns were worried about the safety of the various aging nuclear facilities that dotted the border regions between it and the new Russia. Because the country was concerned about the release of wastewater into the Baltic Sea and the Gulf of Finland, it simply used its own funds to build a wastewater treatment plant for the city of St. Petersburg, Russia. Even though Russia may have had the funds to do this by itself, her reluctance or rather lack of speedy resolution to the shared problems of nuclear waste disposal and effluent control from St. Petersburg had to draw upon Finnish sensitivity to being practical in its approach to problem solving. This is a cultural disposition that is also reflected in the responses that comprise Factor 3 for the construct of traditionalism-identity (Table 8. 3).

Despite its own internal problems, Finland is one of the few countries in the world that oftentimes sees her own problems as harboring similar levels of moral equivalency to that which affects the rest of the world. As a system of basic common values that help shape the behavior of a people in a given socio-political context, cultures have enduring value. In "most preindustrial societies, this value

Table 8.3 Loadings on Third Rotated Factor: Traditionalism-Identity (Culture)

Variables	*Factor*	*Communality*
Accurate Understanding of Problems of Human Society	.535	.302
Technology Can Help Finnish Society to Adapt	.716	.514
Remedies to Social Problems Are Yet to Be Found	.747	.559
Technology Improves Mankind's Understanding	.704	.517
Other People's Ethnic Values Are Just as Good as Mine	.510	.365
Technology Does More Good than Bad	.576	.451
Eigen Value	3.57	
Proportion of Variance	9.93%	
Cronbach's Alpha $\alpha = .763$		

system takes the form of a religion and changes very slowly; but with industrialization and accompanying processes of modernization, these worldviews tend to become more secular, rational, and open to change."[7] Studies by Inglehart and Baker point to the fact that while economic development may be linked to systemic changes in basic values within societies, but there undoubtedly remain "the persistence of distinctive cultural traditions and values that endure despite modernization."[8] If and when value change occurs, it is more as a result of generational replacement as birth cohorts replace each other over time.[9] How has technology changed Finnish culture, and are there any enduring legacies?

8.3 Method and Justification

The survey questionnaire produced a lot of information that can tap into the evolving sentiments regarding various issues of public policy and the challenges facing human society.

A confirmatory approach using the factor analysis procedure was applied to the data as a way of isolating the underlying dimensions and constructs reflected in the initial assumptions and hypotheses.[10] Results from the factor procedure showed that three factors with eigenvalues greater than 1 were extracted by means of principal component analysis using varimax rotation. Principal component extraction is often the preferred method for item reduction and identification of empirically derived domains (dimensions).[11] To identify strong item clusters and consequently cohesive components, a more restrictive scaling construction was used. Items with loadings equal to or less than .45 on any component were dropped from the model, and items with loadings greater than .45 on any of the three domains (dimensions) were included. The factor loadings were found to correspond quite nicely with the hypothesized constructs of social capital, ICT utilization, and traditionalism-identity (residual culture). By this process, the combined independent effects of the various sets of questionnaire responses have been reduced to achieve a rather homogenous effect and elaboration. While the communalities represent the degree to which each of the independent variables (questionnaire responses) explains a portion of the underlying dimension, the total variance explained by all three factor domains was 35%.

To the extent that these represent a grouping of independent responses that have achieved communality through several iterations, what we can learn from the hypothesize constructs is first, that they do exist and second, that it is possible to draw upon these results to confirm the resiliency of tradition and habit, as well as cast novel theoretical explanations and understandings of the dynamics of human action. As a means of further assessing how well the theoretical constructs reflected in each of the extracted components (the factors scores) was supported empirically, the internal reliabilities (consistencies) of the underlying dimensions as reflected in the factors were estimated using the Cronbach's alpha coefficient.[12] "Internal consistency is a measure of general reliability evaluating the extent to which the different items (variables) in a scale measure the same trait."[13] The alpha coefficients (α) for each underlying dimension are presented in Tables 8.1, 8.2 and 8.3. The magnitude of the coefficients point to a strong internal consistency reliability among items in each of the factor domains (*Factor 1*: Social capital, $\alpha > .839$; *Factor 2*: ICT Utilization, $\alpha > .832$; and *Factor 3*: Traditionalism-Identity, $\alpha > .763$).

8.4 In Retrospect

As has been indicated earlier, while national cultures may be resilient to an extent, they are also amenable to change. These changes can come about as a result of increasing levels and nature of economic or political development, advancement in ICT utilization, the structure and systems of education or by means of intergenerational cohort replacements. Cultural change can also come about as a result of global movements due to globalization, commerce, immigration and systemic instability wrought by increasing incidences of institutional decay and moral crisis. In his famed index of national cultures, Geert Hofstede[14] ranked Finland on six cultural characteristics: power distance (33), individualism (63), masculinity/femininity (26), uncertainty avoidance (59), long term orientation (38), and indulgence (57). But for this purpose, the characteristics that are most relevant in this regard are those concerning uncertainty avoidance, masculinity/femininity, and long-term orientation. This study supports the case that these cultural values are still very much evident among the Finns despite the country's advancements in economic development, technology and intensive ICT utilization in both private and public life.

By virtue of the Hofstede Index, Finland is characterized as a feminine society (out of respect and equality considerations for both genders), as a country that has a high preference for avoiding uncertainty, and as a country that is focused on seeking normative and absolute truth and a propensity for quick and immediate results. It could thus be inferred that collectively, these considerations represent an enduring contrast between such values as "equality versus subordination, progressivism versus conservatism, community versus isolation, sympathy versus antipathy, and emotionality versus insensibility."[15] Because people who espouse feminine cultural values tend to be more concerned with maintaining personal relationships, cooperation and interpersonal harmony than those who espouse masculine cultural values,[16] the desire to appear agreeable, the concern for harmonious social relationships, their greater expressiveness and the concern for the socio-emotional wellbeing of others endear them to greater influencability.[17] To the extent that much of these can be nuanced from the factor loadings in Tables 8.1, 8.2 and 8.3, there is no doubt that while aspects of Finnish culture may have changed or at best been modulated, the country still retains its foundational cultural attributes. And these cultural attributes and traditions seem to have endured despite modernity and economic development.

While this is not to suggest that its cultural values have remained intact over time, the country has become more realistic about the state of the world and the place of human society in it. Its cultural orientation has become less insular but more cosmopolitan and outward-looking beyond its natural borders. One can thus surmise that this renascent cultural disposition transcends an evolving extra-national (global) cosmopolitanism on an international scale. It therefore sees technology as a tool for understanding the realities of the modern world and as a means of adapting to its multiple and disparate challenges. In the same way that technology helps to reinforce aspects of Finnish culture (such as their 'can do' attitude), Finnish culture is used to reinforce technology or ICT utilization—both becoming more and more embedded into each other. While it has been argued that there is a "relationship between cultural individualism and economic affluence; it is also equally stated that countries that became wealthier also seem to have become more individualistic—a process visible in many Asian countries like Japan, South Korea, and Thailand"[18] And for Finland, there seems to be a bi-directional relationship between technology and culture—mediated only by that enduring and legendary Finnish passion—*Sisu.*

But I must also be quick to point out that this observation is quite different from the country's history of international outreach and involvement in crisis or conflict negotiations. That activity is driven more by a political as opposed to a cultural logic. But in the midst of all these, there are stark questions to be asked: What will happen to the Finland of tomorrow, and what will become of its Nordic identity? Will the current generation carry on with the cultural and resilient ethos of the past, and what will be history's final verdict? What does EU membership mean for Finnish identity? What is the difference between Finland as a Nordic country and Finland as European, and to what extent could they be mutually exclusive? What about immigration and asylum seekers, particularly Somalian refugees and the Middle-Eastern immigrants? Will they ever become Finnish and how, or would they portend an ethno-religious powder-keg waiting to be ignited? To what extent would the dominant Lutheran religion be affected? The seemingly and oft-spoken taciturn nature of Finnish life may be masking a deep-seated concern for what the future will hold and whether holding on to what has always been Finnish would matter after all to the new generation. As Finland advances to the top of all major global indicators of growth and development, education, science and technology, unbeknownst to many of its citizens, its socio-cultural infrastructure may be unraveling at the seams in a fashion reminiscent of divergence and integration both occurring at the same time.

Notes

1 Delbridge, R. and T. Edwards. 2013. Inhabiting institutions: Critical realist refinements to understanding institutional complexity and change. *Organization Studies,* 34(7): 927–947, p. 935.

2 Coleman, J. S. 1990. *Foundations of social theory*. Cambridge, UK: Belknap Press.

3 Putnam, R. 1993. *Making democracy work: Civic traditions in modern Italy*. Princeton, NJ: Princeton University Press. Also see R. Putnam. 1995. Bowling alone: America's declining social capital. *Journal of Democracy, 6*: 65–78.

4 Putnam, R. 1995. Bowling alone: America's declining social capital.

5 Kavanaugh, A. 2001. Civic engagement among community computer network users: Trend or phrase? Research conference on communication, information, and Internet policy. Alexandria, VA: Telecommunications Policy Research Conference.

6 Haythornthwaite, C. 2005. Social networks and Internet connectivity effects. *Information, Communications & Society,* 8: 125–147. Also see Kalu N. K. and B. Remkus. 2010. The evolution of social capital and civic engagement between nonprofit networks and county representatives: A social constructivist approach. *Social Science Computer Review,* 28(1): 135–150.

7 Granato, J., R. Inglehart, and D. Lange. 1996. The effects of cultural values on economic development: Theory, hypotheses, and some empirical tests. *American Journal of Political Science,* 40(3): 607–631, p. 608.

8 Inglehart, R. and W. E. Baker. 2000. Modernization, cultural change, and the persistence of traditional values. *American Sociological Review,* 65: 19–51, p. 19.

9 Inglehart, R. and P. R. Abramson. 1994. Economic security and value change. *American Political Science Review,* 88(2): 336–354, p. 339.

10 Gregg, P. M., and A. S. Banks. 1965. Dimensions of political systems: Factor analysis of a cross-polity survey. *American Political Science Review,* 59: 602–614. Also see J. O. Kim and C. W. Mueller. 1978. *Introduction to factor analysis: What it is and how to do it.* Newbury Park, CA: Sage; and R. J. Rummel. 1967. Understanding factor analysis. *The Journal of Conflict Resolution,* 11: 444–480.

11 Nunnally, J., and H. Bernstein .1994. *Psychometric theory*, 3rd ed. New York, NY: McGraw-Hill. Also see T. J. Beebe et al. 2001. The community readiness survey: Development and initial validation. *Evaluation Review,* 25(1): 55–71, p. 61.

12 Cronbach L. 1951. Coefficient alpha and the internal structure of tests. *Psychometrika,* 44: 373–393.

13 Beebe, T. J., P. A. Harrison, A. Sharma, and S. Hedger. 2001. The community readiness survey: development and initial validation, pp. 55–71.

14 http://geert-hofstede.com/finland.html. Accessed on November 8, 2015.

15 Leidner, D. E. and T. Kayworth. 2006. A review of culture in information systems research: Toward a theory of information technology culture conflict. *MIS Quarterly,* 30(2): 357–399, p. 371.

16 Hofstede, G. 1984. *Culture's consequences*. Newbury Park, CA: Sage Publications.

17 Eagly, A.H. 1978. Sex differences in influencability. *Psychological Bulletin,* 85: 86–116. Also see Venkatesh V., M. G. Morris, P. Ackerman, and T. Sykes. 2004. Individual reactions to new technologies in the workplace: The role of gender as a psychological construct. *Journal of Applied Social Psychology,* 34(3): 445–467.

18 Hofstede, G .1999. Problems remain, but theories will change: The universal and the specific in 21st –century global management. *Organizational Dynamics,* 28(1): 34–44, p. 39.

9 Of Black Swans and the Human Condition

The Committee for the Future

There are certain things that people think about but quickly set aside simply because the probability of that event coming to pass is so remote that it would not be worth placing a bet on it. The reality is that even though we live a life of great promise and also make plans for tomorrow, we live and plan not with the benefit of certainty of fact but in anticipation because the future can always escape our imagination. "The necessity to live with profound uncertainties is a quintessential condition of our species,"[1] so we live but precariously. This is evidenced by the state of the world during the Cold War years (1945–1991), which ended following the disintegration of the Soviet Union in 1991. But for much of the 45-year period, the fate of the world relied essentially on a deterrence framework anchored on an imagined reality conceived in the form of *mutually assured destruction*, otherwise known in strategic parlance as the MAD Doctrine. In line with the natural instinct of all humans for survival and as rational actors, states understandably also seek survivability; hence, the issue of reciprocal or equal opportunity suicide never materialized. But does that then diminish the possibility?

9.1 The Possibility of the Impossible: What If?

What if the impossible should become possible? What if the unimaginable becomes imaginable? What if the world runs out of oil, and there is no more energy to fuel our electricity? To be more practical, what if the earth shuts down and falls out of orbit? What if the sun never sets and human life becomes threatened? What if a mega-sized meteor falls into the ocean and causes massive flood and drowning of low-lying communities and regions of the world? What would happen if Iran gets a nuclear bomb? What if Russia and the United States engage each other in a nuclear Armageddon? What would become of the world? What if a sustained nuclear exchange shakes the earth with enough intensity and jolt to result in a nuclear winter? A nuclear winter is a hypothetical climatic effect in which a nuclear firestorm incinerates cities and forests, leaving torrents of smoke that ascend into the atmosphere and entomb the planet in a billowing, black cloud of ash—that results in noontime darkness, plummeting temperatures and the eventual death of life on planet earth.[2]

The above scenario is the stuff that make *black swans* a fitting metaphor that could help to illuminate the increasing randomness that often manifest in rare surprising but unexpected events like 9/11. These events generally represent large deviations from the norm that oftentimes cast a dark shadow over the affairs of the world we live in. The theory of black swans is attributed to Nassim Nicholas Taleb in his much-acclaimed book *The Black Swan: The Impact of the Highly Improbable*,[3] in which he refers to unexpected events of large magnitude and consequence and how such events, often considered extreme outliers, collectively play vastly larger roles than regular occurrences in history. Black Swans[4] are the unlikely and unforeseen events that change the world. While they seem to come out of nowhere (like the 9/11 event), mathematical probability can rarely predict them, and conventional wisdom is blind to them. At various stages in their evolution, black swans may be reflected in an evolving pattern of continuities and discontinuities, development of weak signals, megatrends and problematic wild cards that, though unsuspectingly, mask a more ominous outcome.

But it should be understood that "black swans and the effects they produce may be either negative or positive in nature; such that events that change the world and society may occur in any area at all: nature, technology, economics, politics, culture, values, beliefs, customs and habits."[5] For instance, "humans could achieve eternal life by slowing down cellular metabolism so that our organs would remain forever young by means of regeneration";[6] as 3-D printers become more ubiquitous, they may become a *black swan* that revolutionizes the manufacturing industry beyond what has hitherto been contemplated. We might reach a level of discovery that "besides utensils, furniture, food and electronics, we might also be able to print replacement parts for the human body."[7]

The "bolder and more long-ranging our investigations of alternative paths and events for the future are, even those that might seem virtually impossible; the greater will be our competencies and ability to prepare for the future while minimizing risks and proactively seizing opportunities."[8] But above all, one thing we must remember is that while unintentional black swans are rare, the majority of them are either related to upheavals in the natural world or events or sequences of events catalyzed by people's own actions.[9] We have very little control regarding when and how they would occur and the scope of intensity or lack thereof should their eventuality become most evident. Reliance on predictions are therefore an expedient means for guiding public policy while at the same time, but to a certain degree, limiting the odds of total failure in the midst of overwhelming surprises.

9.1.2 System Under Duress: Continuity and Discontinuity

There is an enduring lesson to be gleaned from any study of systems theory. The inter-relationship of parts in terms of their distance and location from each other and their respective functions or roles relative to the others produces a structural-functional framework needed to achieve a specific goal or objective. To survive and endure, the framework must at the same time be capable of resisting the centrifugal forces of inertia as well as the draw-down effects of negative entropy; and to do this, it must maintain homeostatic equilibrium, otherwise known as stable states. Because all systems, whether obvious or not, always gravitate toward dissolution, the deliberate effort to arrest the entropic process is a continuous yet unpredictable one; even if systems are not subject to externally induced disruptions, there are other internal forces that can overtime create conditions for system decay.

In his *Global Catastrophes and Trends: The Next 50 Years*, Vaclav Smil paints a rather vivid portrait of the United States in a slow but inexorable decline. He argues that this complex process of decline has been ongoing unbeknownst for several years but has suddenly become more prominent and profound, "coalescing into a new amalgam of worrisome indicators that point unequivocally toward a gradual decline."[10] In so far as systems are exposed to various environmental factors, existing equilibrium states can be punctuated not only by gradual incremental changes but also by sudden bursts of fundamental change as radical as they are transformative. Fundamental change creates surprises, not because something dramatic has happened, but because of the suddenness, scope and the unexpected nature of the change.

This has been the way of the world. The global system of states has always been in a state of crisis, from one discontinuity to the other, and from one brief moment of relative peace to a state of cataclysmic disorder. A cursory glance at John G. Stoessinger's *Why Nations Go to War*[11] would demonstrate that conflict and crises have always been the way of the world, and the brief equilibrium states achieved over the years can best be summarized as an active volcano that had lain silent for some time. As the heat energy percolates and bubbles underground without any obvious outlet; and at the final moment, its anger is unleashed unto the world in an uncontrollable fury that, more often than not, leave a trail of death and devastation in its aftermath. The volcanic analogy as utilized here also helps to illuminate the fact that similarly situated seething undercurrents of energy have oftentimes exploded unto the international scene in ways that were previously unimaginable.

The Spread Effects of Small Beginnings

Without making any forays into ancient history, examples abound of a world full of surprises bent on its own self-destruction: such as the sudden invasion of the Soviet Union on June 22, 1941 by Nazi Germany under Operation Barbarossa (1941–1942), the German invasion of Czechoslovakia and the consequent occupation of the Sudetenland under the Munich Agreement of September 30, 1938, the dropping of the first atomic bomb on Hiroshima (August 6, 1945) and Nagasaki (August 9, 1945) that quickly brought Imperial Japan to its knees, and the Korean War (1950–1953)—a tinderbox that remains with us today and may rekindle at any opportune moment. In the larger picture, there was a World War I (July 28, 1914–November 11, 1918) and World War II (1939–1945) caused by several underlying political and territorial issues among the great European powers that otherwise culminated in the ill-fated assassination of Archduke Franz Ferdinand (heir to the Austro-Hungarian throne). The ensuing conflagration consumed millions of lives by the time it had rested. Other more contemporary but equally dramatic events that also resonated unto the international system included the fall of the Berlin Wall (1989), the disintegration of the Soviet Union (1991) and the 9/11 attacks of 2001.

Today, the fundamentals of a global system under stress have acquired a modern dimension of an old story: narco-trafficking, human trafficking, regional insurgencies and civil wars, the rise of religious fundamentalism and its overt manifestation in global terrorism. The world has experienced terrorism for much of its checkered history, but today's version has become more virulent and sophisticated. Many more countries have acquired the nuclear bomb since it was first used in 1945, and many others are vehemently seeking it. Ironically, many of these countries claim to be seeking nuclear power development for *peaceful* and *civilian* purposes. As Robert Dahl cautions us in his *Controlling Nuclear Weapons: Democracy versus Guardianship*[12] to the effect that while "technology is often neutral, the purposes to which it is put are not; hence we have little ground to assume that they will be used for benign purposes, unless we deliberately shape the institutions that determine their use." Today, anyone who has any feeling for the ever-deceptive volatility of the Middle East would see that the black swans, rather inexorably, are continuously circling like vultures poised to burst into the open from anywhere at any moment in time.[13]

On the *remote* possibility that all countries that seek to obtain a nuclear bomb would eventually get it, the question then becomes: What if some of them actually use it? How can societies survive and remain functional in the aftermath? Can such a catastrophe ever be contemplated? Even though it may not perhaps come to be, the Finns have found a way to address similar questions of the unknown. While we cannot predict what the future will be with certainty; "it can, in many significant ways, be imagined and invented."[14] And as such, we often anchor our decisions on the basis of probabilities. But regardless of whether an event has a 5%, 10% or 50% chance of occurring, that there is a probability of its occurrence in the first place is all the more reason we should develop policies to safeguard against it as if it were inevitable. This is part of the driving logic that informs the activities of the Finnish Parliament's Committee for the Future.

9.2 The Committee for the Future: The Role of Futures Studies in Public Policy

The Committee for the Future is one of 16 Standing Committees of Finland's Parliament (Eduskunta) and is charged with dealing with, among other things, long-term issues concerning sustainable development for the future, Parliament's oversight role on research and societal implications of technological development, studies of future trends and transformations, research and methodological assistance to other parliamentary Committees on future-related issues and reports on long-term effects of Parliamentary decisions. It seeks to address important existential questions concerning human challenges and survival in the future society. In addition to proffering solutions for understanding and strengthening the ecological environment, it also seeks possible answers geared toward ensuring that human society can continue to function even after exceptional or traumatic events such as black swans. In the words of Päivi Lipponen,[15] What would the world and life be like in the future if a *Black Swan* comes to pass?

The Committee is made up of 17 members drawn from the different political parties represented in the Parliament and is the "only forum in Parliament where all parties can collectively appraise the development of the entire political system against a time frame that is longer than the conventional day-to-day demands of legislative politics."[16] It was established in 1993 and, in collaboration with the Finland Futures Research Center in Turku University, has been involved in developing a national *foresight system*. There is also increased cooperation with other agencies such as the Academy of Finland, Tekes (The Finnish Funding Agency for Technology and Innovation) and Sitra (The Finnish Innovation Fund) as well as other international collaborative engagements on futures studies. The collaborative effort[17] is based on a general recognition that we can draw upon the role of science, technology and the creation of new concepts and ideas in addressing problems of revitalizing existing institutions as well as thinking about the things that would be reflected as permanent archetypes of the future and what would not be. While the job of the Committee is not to do research on its own, its activities revolve around leveraging the results of futures research in making policies that address issues of the future.

Epistemology of the Unknown

In *Revamping the Work of the Committee for the Future: Summary*, Paula Tiihonen briefly delineates the origins of futures studies.[18] She points out that, although it may be difficult to pin-point the exact date, the general consensus has been that the scientific study of futures began soon after WW II, around 1946 when the German professor Ossip K. Flechtheim framed the concept known as futurology. But references to the insight into futures studies have been attributed to the earlier works of the famed English writer and novelist Hebert George Wells (1866–1946) who first advocated for futures studies (foresight) in a lecture he delivered in 1902. Wells is credited with such future-oriented works like *The World Set Free*, *The Time Machine*, *The Invisible Man* and *The Shape of Things to Come*.

However, "following the stock market crash of the 1930s and the lessons of W. W. II, futures studies became institutionalized as a part of social sciences research and political decision making."[19] But it was in the 1950s and early 1960s that futures research began to gain greater currency in Europe and elsewhere. The 1960s particularly was a period of scholarly ferment that saw the emergence of futurists like Herman Kahn, Bertrand de Jouvenel and the Hungarian-British scientist Dennis Gabor. And this tradition has continued among European and American scholars, research and military institutions to date.

While the Committee for the Future[20] does not generally deal with legislative proposals, it has the power to decide its own agenda and to tackle proposals and projects of its own initiative. Its "primary task is to think about and work towards the best possible future for the people of Finland."[21] In this regard, it seeks to deal with a broadened array "of issues and their long-term implications, propose different alternatives while highlighting the dangers, and to build competing scenarios that draw upon existing methodologies in futures research."[22] Its ideas generally find their way into the *Government's Programme for the Future* (to be submitted at least once in every electoral period or in a four-year parliamentary cycle) and are deliberated in plenary sessions of Parliament. The Committee serves as Parliament's think tank by helping to bridge the gap between research and corresponding decision making, creating increased awareness about the future while educating the public at large on the social obligation and sacrifices that will be required.

Among the areas in which the Committee has done work include but are not limited to issues concerning sustainable growth, an inspired society, the means for acquiring new knowledge and competencies needed to function effectively in the society of tomorrow, evolving demographic trends, population policy and preparations for changes in a new age structure, biopolicy and issues of climate change, the social and human effects of increasing urbanization, the long-term fate of the welfare state and its sustainability, the future of democracy and politics as system-level governance mechanisms, crowdsourcing for legislative decision making and the quest for radical technologies that would change the world as we know it. Overall, the Committee sustains a chastened view of the role of Finland and its capacity to adapt and remain vibrant in light of the enormity of the challenges we face in an ever-changing world.

9.2.1 Radical Technologies: Agenda for the Future

What would the society of the future look like? How would humanity be able to adapt to its many challenges? What would be the role of technology in this regard? The Finns have already begun to think and reflect upon these questions not because of any likelihood that future events will unfold in some specific ways; but because when we fail to challenge the limits of human imagination, the only thing that can escape us is its hidden potentials and possibilities. In a world fraught with uncertainties much of which remain unknown to us, we relng on probabilities as a means of reducing the odds that one strike would mean that we are out. But the Finnish approach is not only about finding ways to deal with or adapt to unknown cataclysms such as black swans; it is equally about finding social, environmental, and medical solutions to contemporary and future problems of society, exploring their possibilities and seeking novel and revolutionary ways to understand and deal with them. This ability, therefore, offers a template for "coping with assorted crisis and to transform many events from potentially crippling milestones to resolved challenges."[23] Through continuous innovation and cutting-edge research projects, Finland has remained at the forefront of a developing scientific genre generally referred to as *radical technologies*.

The premise of radical technologies is based on the need to provide a comprehensive assessment of the *future of technology* and the *technology of the future* including a review of weak signals and disruptions already evident in existing technological applications. Hence, an effective understanding of the technology of the future premised "on a long-term general analysis for Finland would help the country in preparing a futures roadmap for use in a global, unpredictably changing world that contains new and multidimensional uncertainties."[24] On a more continental scale, the ideal of radical technologies fits in properly within a broader European *Future and Emerging Technologies* (FET) framework[25] aimed at turning Europe's excellent science base into a competitive advantage. By reinforcing unexplored collaborations between advanced multidisciplinary science and cutting-edge engineering, FET projects stand to open up promising avenues toward powerful new technologies beyond what is currently known.

The Committee for the Future[26] is well-acknowledged for its work in collating opportunities, threats and weak signals relating to the future. The collective image is one of people who take delight in getting things done, producing good ideas and passing them on to other Committees as input in their legislative decision makings. But its work goes beyond such seeming simplifications. In 2015, it released its seminal work titled *100 Opportunities for Finland and the World*, in which the Committee brought together a series of innovative and radical technologies that, would address various multiple needs of contemporary society but also offer new opportunities for attending to the problems of the future. In order to qualify as a radical technology and be included in the list of 100, the technological solution has to be relevant, or at least crucial, for many global value-producing networks.

While "a radical technology implies that a technological solution would be able to change current practices either by saving costs, easing people's everyday lives, the minimum requirement is that the principles upon which the technological solution is based must already have been proven in a scientific publication."[27] They should be ones that promise to offer actual solutions as well as have a broader impact on the market and potentially the global scene. They must be functional in serving the needs of the market or the society. In addition to listing the most probable areas of application, radical technologies are expected to deliver on their promise within a stipulated timeline, and to demonstrate realistic prospects in different value-producing networks or markets by at least between 2020 and 2030.

The published list includes technological solutions offered in such complex areas as human-machine interface technologies, routine and complete DNA sequencing, biochips and biosensors for self-diagnoses, portable magnetic resonance imaging scanners, drugs from genetically modified organisms, nanorobots in health promotion, longer lifetime and slower aging processes, brain implant that restores or develops brain functions, artificial muscles, artificial self-renewing skins, drugs that prevent dementia, cell cultivation for the repairing and regrowing of human organs, synthetic cartilage for human joints, 3-D printing of human organs to replace animal testing, cyber insects, schools in the cloud (in the form of computer-based collaborative learning), self-driving cars, emotion management in robots and automatic recognition of emotions, quantum phenomena, human recognition systems, continuously monitored personal health, freely organizing remote work and organizations that form on the Internet, and many more.

In a nuanced way, radical technologies are no more in the future; they are already here, and most of them will be in the market or in use within the next 15–20 years if not earlier. The role of Finnish research centers, universities and private enterprises in spearheading this effort has been exemplary. In the same way that social evolution is not a static phenomenon, and as most existing technologies reach maturation point, human survivability on earth would depend on finding new ways and novel approaches to dealing with everyday activities. Radical technologies may offer new pathways to human adaptation, but they will also bring about dramatic changes to the existing social institutions and structure, the character of relationships between citizens and the central role of government in devising policies that address issues of the common good. Better than most countries, Finland understands the trajectory in which technological innovation is poised to drive human society. Despite its size, the country carries on its shoulders a greater proportion of the burden in the perennial search for human solutions regarding how we live and survive in a finite world. But at some point in the future and like all other things before it, technology will have run its course. And the most pressing question then for future futurists would be: After technology, then what?

Notes

1 Smil, V. 2008. *Global catastrophes and trends: The next 50 years*. Cambridge, MA: MIT Press, p. 246.
2 Lamb, R. 2009. What would nuclear winter be like? January 13, 2009. HowStuffWorks. com. http://science.howstuffworks.com/nuclear-winter.htm. Accessed on August 10, 2015.
3 Taleb, N. N. 2007. *The black swan: The impact of the highly improbable*. New York, NY: Random House, Inc.
4 Dickey, C. 2013. Iranian Bombs and Black Swans. *The Daily Beast*, December 17, 2013. http://news.yahoo.com/iranian-bombs-black-swans-104500828–politics.html. Accessed on August 10, 2015.
5 The Committee for the Future. 2013. Black swans: What will change the world next? Helsinki, Finland: The Finnish Parliament's Committee for the Future, pp. 22–23.
6 Ibid. p. 30.
7 Ibid. p. 36.
8 Ibid. p. 39.
9 Ibid. p. 29.
10 Smil, V. *Global catastrophes and trends: The next 50 years*, p. 141.
11 Stoessinger, J. G. 2010. *Why nations go to war*. Belmont, CA: Wadsworth/Thomson Learning.
12 Dahl, R. 1985. *Controlling nuclear weapons: Democracy versus guardianship*. Syracuse, NY: Syracuse University Press.
13 Dickey, C. Iranian Bombs and Black Swans, p. 1. http://news.yahoo.com/iranian-bombs-black-swans-104500828–politics.html. Accessed on August 10, 2015.
14 Dator, J. 2007. Governing the futures: Dream or survival societies? In *Human competency development in the information society*, ed. J. Kasvi, P. Tiihonen, and K. Nolvi, 15–24. Helsinki Finland: Proceedings of the Fifth General Assembly of the International Parliamentarians' Association for Information Technology (IPAIT V), Committee for the Future, Parliament of Finland, p. 20.
15 Lipponen, P. 2012. The Committee of the Future is one of a kind. Paper presented at The 28th IFLA Pre-Conference of Library and Research Services for Parliaments, Helsinki, Finland, August 9, 2012 (Päivi Lipponen is a member of the Finnish Parliament and Chair of the Committee for the Future).

16 Tiihonen, P. 2014. A committee for the future. *Future Justice*, February 17, p. 3. www.futurejustice.org/blog/guest-contribution/guest-article-a-committee-for-the-future. Accessed on July 24, 2015.
17 Lipponen, P. The Committee of the Future is one of a kind. (Päivi Lipponen is a member of the Finnish Parliament and Chair of the Committee for the Future).
18 Tiihonen, P. 2011. *Revamping the work of the Committee for the Future: Summary*. Helsinki, Finland: Committee for the Future, Parliament of Finland, p. 6.
19 Ibid. p. 6.
20 Tiihonen, P. A committee for the future, p. 3. www.futurejustice.org/blog/guest-contribution/guest-article-a-committee-for-the-future. Accessed on July 24, 2015.
21 Ibid. p. 1.
22 Ibid. pp. 2–3.
23 Smil, V. *Global catastrophes and trends: The next 50 years*, p. 251.
24 Lipponen, P. The Committee of the Future is one of a kind, p. 4. (Päivi Lipponen is a member of the Finnish Parliament and Chair of the Committee for the Future).
25 European Commission, http://ec.europa.eu/programmes/horizon2020/en/h2020-section/future-and-emerging-technologies. Accessed on July 24, 2015.
26 Eduskunta (Parliament of Finland). 2012. The committee for the Future's vision, strategy and operational plan for 2011–15. *Draft*, April 17, p. 3.
27 The Committee for the Future. 2015. *100 opportunities for Finland and the world: Radical technology inquirer (RTI) for anticipation/evaluation of technological breakthroughs*. Helsinki, Finland: Committee for the Future, The Parliament of Finland, p. 49.

10 Finlandia

'Our Land' and a Postscript

In many respects, Finland is very much a land of contrasts. It is one whose history and experience has always served as an enduring but unifying force upon which a sense of pride for country and citizen has remained high among the people. While love of country is a cherished ideal among all national cultures, recent history has shown that sooner than later, the ideal succumbs to the allure of perverse nationalism by which one national culture is extolled as being superior to that of others. But the Finnish love of country is different; it is a sentiment driven by a keen appreciation of the country's uniqueness, in and of itself, in relation (but not in competition) to the experience and history of others. In a nuanced way, this sentiment is also reflected in Finland's national anthem *Maamme* (Finnish) or *Vårt Land* (Swedish) or *Our Land* (English) in the sense that it shows by its own lyrics, a collective admiration for the land (not the people) and what it offers for the sake of its inspiring beauty and existence. But if, on the contrary, then it would only be to the extent that *land, people* and *culture* are construed as integral and synonymous with *Our Land.*[1]

Hence, an understanding of Finland must perforce include a broader appreciation of its people, their history, orientations, how they live and the meanings and emotions they attach to what they do. In addition to their characteristic disposition toward *sisu* (courage and perseverance), the Finns are a people driven to festivities all year round, particularly during the summer months. There are festivities such as the *Midnight Sun Film Festival, Jutajaiset Folklore Festival, Kalott Jazz and Blues, Oulu August Festival, Suonenjoki Strawberry Carnival, Kaustinen Folk Music Festival, Ilosaari Rock, Full Moon Dance Festival, Tuska Open Air Metal Festival, Kihveli Soikoon Skiffle Music Festival, Flow Festival, Down by the Laituri* and many others. In the Laplands, the nightless nights of the North set the scene for music of all genres, as well as movie screenings, art exhibitions, and dance celebrations.[2] In the South, the all-Finnish Viapori Jazz Festival sets alight the picturesque island of Suomenlinna (just off the coast of Helsinki) every August with jazz music of all notes and denominations. The credit for the resurgence of "Finnish jazz can be traced directly to the country's institutes of higher education—the Helsinki Pop & Jazz Conservatory which added jazz to its repertoire in 1972; and the Sibelius Academy which introduced formal jazz studies in its syllabus in 1983."

While summer vacations and festivities offer the average citizen a physical break from work, its more cultural appeal lay in the fact that it provides an essential opportunity to return to nature or to the woods as a way of connecting and reflecting on the country's beginnings. The woods also have become a place of refuge for urban professionals and family members seeking respite and solitude from the hustle and bustle of everyday cosmopolitan life. As exemplified by Kullervo in Elias Lönnrot's *The Kalevala*:[3]

> *Homeward others go, to their dwelling they travel. I have my home in the wilds, on the heath my farm; out in the wind my fireplace, and in the rain my sauna steam.*

The woods and the outdoors have a special placed in Finnish life. It would thus suffice to state that the country's oftentimes harsh and frozen climate has also, in fundamental ways, shaped most of the outdoor festivities and recreations like cross-country skiing, ice fishing, snow shoeing, dog sledding, snow mobiling, sauna, the Midsummer (summer solstice) celebrations known as the *Juhannus*, the Christmas celebrations and watching the northern lights (*Aurora Borealis*) play its magical wonders.

There are other recreational activities or events in Finland that most other cultures would find intriguing such as the wife-carrying championships, the ant-nest sitting championship, world championship for mosquito killing, mobile phone throwing championship or the swamp soccer championships and celebrations every October 13th marking the *Day of Failure*—celebrating shortcomings and failure, sharing experiences as such and promoting the idea and understanding of failure as a learning experience.[4] To the extent that they bring families, friends and fellow community members together, cultural activities are seen as a civic way of life that helps to sustain such social values as creativity, education and shared experiences.

10.1 Culture as Myth and Ceremony: An Evolving Milieu

In many ways, age, physical handicap and a person's economic status do not present limiting factors to the experience of Finnish culture. "The whole cultural experience is heavily subsidized by the state and the municipalities thus making it possible for many people to be able to attend concerts, theatre, dance classes and other festivals."[5] Children's culture is one of the most developed in the world with a network of centers spread throughout the country. "There are also several children's festivals focused on literature, movies, theatre; and the early emphasis in the arts in *basic education* has also helped to nurture and produce a surprising number of world class musicians, singers, and conductors from Finland."[6]

Cultural practices serve not only as avenues for enjoyment and critical reflection on the life and history of a people; they also reflect moments of creativity and public education. The use of oral history, which included *Karelian* rune songs, poems, stories, folklores, fables, superstitions and myths, was central to the transmission of Finnish cultural heritage and life experiences from generation to generation. But while it could be said that "many other factors contributed to the political and artistic renaissance of Finnish culture and identity, Lönnrot's

The Kalevala (Land of the Heroes) was by far the most celebrated element of change that also played a vital role in fostering Finnish national consciousness and identity."[7] In the same way that oral history has played a central role in the development of Finnish culture, so have food, agriculture and the people's traditional attachment to the land and forest. While some family farms remain in several rural areas of the country, "increased mechanization and specialization in farm production (diary, cattle, hogs and grains) which began in the 1960s has led to a dramatic shift in the labor force to manufacturing and the service sectors."[8] Today "less than 11% of the labor force is involved in agriculture and forestry."[9] For this reason, people go out of their way to transform some of the most unusual flora and fauna into edible great food.

Among the categories of culturally significant food and drink are the basic sausage (*makkara*), cheese bread, yeast coffee bread (*pulla*), brown malted porridge (a traditional Easter diet known as *mämmi*), mead (*sima*), potato pasties (*perunapiirakka*), homemade beer (*talouskalja*), soured whole milk (*viilia*), blood sausage (*veren makkara*), smoked fish, and jerked reindeer meat. Some of the food items are regional specialties served either commonly or during special festivities and occasions. But as they say, you know a people by what they eat; for the Finns, food is an integral part of their culture. The "Finns are also known to have the highest consumption of coffee in the world; consuming approximately 12 kilos per person per year."[10] Across all age demographics, about 14% of men and 6% of women reported consuming more than10 cups of coffee per day."[11] It is considered a national drink celebrated on specific occasions as a mark of grace and hospitality, and it is liberally consumed across the country.

10.1.1 Of Theatre, Music and the Arts

Finland has contributed immensely to the development of visual, performance and graphic arts. Much of the artistic expressions draw inspiration from the forces, shapes, colors and textures of the northern landscape and the human relationship to nature.[12] These influences have found their way into various paintings, sculpture and other art forms, some critical but others extolling the travails and triumphs of the Finns. Finland has nearly 30 symphony orchestras and more than a dozen major festivals offering classical, folk and contemporary music; there are about 60 institutional theatres, mostly in cities and large towns, and some outdoor summer theatres, which are often operated to correspond with community festivals.[13] The "annual festivals of various arts, including opera held at places such as Helsinki, Vaasa, Kaustinen and Savonlinna and at various museums show an awareness of the individuality and importance of Finland's contribution to world culture."[14]

The country has "more orchestras, music institutes, theatres, museums and festivals per capita than any other country in the world; and although the government plays a central role in arts policy, arts institutions are not created under centralized state control but have generally been initiated at the local levels."[15] It is only after the institutions have proven to be viable as well as to command future prospect that state support becomes readily available. "Since 1969, the state has administered a system of artist grants that allocate a tax-free monthly stipend over a specified period to artists working in architecture, motion pictures, crafts, design, dance, literature, music, theatre, and visual arts."[16] The productions that result from these help to reinforce national identity and pride, as well as increased opportunity for the international export of Finland's creative potential.

The Finnish National Opera in Helsinki, the Savonlinna Opera Festival, the Sibelius Academy, the Helsinki Philharmonic Orchestra, Helsinki's Ateneum Art Museum and the Finnish Radio Symphony have offered varied opportunities to enjoy top class music and arts and opera from such Finnish singers and composers such as Karita Mattila, Jorma Hynninen, Soilie Isokoski, Joonas Kokkonen, Magnus Lindberg, Kaija Saariaho[17] and many others. Although state subsidies fund a majority of the arts institutions, the proactive emphasis on the arts and the compulsory requirement for music education particularly at the basic school level has generated a lot of national talent. Today, Finland prides itself on having produced performers in various genres of the arts, opera singers, composers and conductors, many of whom have enjoyed prestigious careers across Europe and other parts of the world.

One of the greatest Finnish composers was Jean Sibelius (1865–1957), whose works achieved international renown including such timeless themes as *Kullervo* and *Tapiola*, as well as the epic song the *Finlandia* (see YouTube).[18] It has remained a source of national pride and Finnish identity. The *Finlandia* also spread its magical allure to the erstwhile Republic of Biafra when it was chosen as the national anthem of the short-lived state, which had unsuccessfully seceded from Nigeria (1967–1970). The hymn was composed by Sibelius in the autumn of 1899 as a series of illustrations depicting Finland's past, as well as its resistance to the increasingly restrictive Russian rule at a time when the country served as an autonomous Grand Duchy of the Russian Empire.[19] However, it was not sung in Finland until Finnish words for it were written by the opera singer Wäinö Sola in 1937.[20] It was only after the Winter War with the Soviet Union (1939–1940) that the Finnish poet V. A. Koskenniemi provided a new text for the *Finlandia*, which has persisted till today.

Sibelius's birthday of December 8 is celebrated in Finland and internationally; the country has done a lot to immortalize his memory and contribution to Finnish and worldwide symphony and music. Every aspect of Sibelius's life was typically Finnish, from his early humble beginnings in the southern Finland town of Hämeenlinna to the capital Helsinki, and in the midst of his international travels as a composer, he still found solace and rest in rural suburbia. His home in Ainola (near Lake Tuusula, Järvenpää) is a national museum (opened in 1974), and there are also the Sibelius Monument and the Sibelius Academy, both in Helsinki, and a museum in Turku all of which continue to serve as great tourist attractions. Like most nations, Finland has a devotional habit of honoring its heroes with statues and monuments.

10.1.2 Literary Activism as Civic Duty

Finnish literature, in the form of folklore, narrative, poetry, fictions and story novels, can best be explained as eclectic, often expressive, allegorical, abstract and emotional. Drawn from a major influence of Swedish and Russo-Karelian underpinnings, contemporary Finnish literature seeks to combine historical narratives, post-structuralist and modernist expressions as a means of interpretation or providing explanations about the human condition. While literature offers a bridge between language and identity, it also serves as an outlet for refocusing the public debate on the role of history and experience in shaping the prospects and fortunes of the future. Following in this tradition was Finland's national writer Aleksis (Stenvall) Kivi (1834–1872) through his many works in poetry, plays and literature.

While much of the early Finnish literature was published in Swedish, particularly the works of the national poet Johan Ludvig Runeberg (1804–1877), Zachris Topelius (1818–1898) and the latter works of Edith Södergran, and Tove Jansson; but Kivi's epic work *The Seven Brothers* (1870) is considered to be the first novel written in the Finnish language. The novel "tells the story of some freedom-loving village youths who took to the woods and lived a life of adventure but gradually matured and finally accepted the responsibilities of becoming sober citizens."[21] Mixing elements of realism, nostalgia and humor, the work illuminates the constant struggle between man and nature, good and evil, between pristine innocence and repudiation of guilt in a renewed affirmation of the self. While it helped to inspire a growing national consciousness and identity, it also became a source of motivation that led to creative developments in other aspects of the arts.

Despite neoromanticist leanings, a strong realist current seems to have permeated Finnish literature beginning in the early twentieth century, as exemplified in Minna Canth's depictions of women and the poor, and Juhani Aho's treatment of emotional experience.[22] Besides Elias Lönnrot's *The Kalevala*, other Finnish writers and poets of different genres include Mika Waltari (*The Egyptian,* 1945), Väinö Linna (*The Unknown Soldier,* 1954), Mikael Agricola, Eino Leino and Frans Eemil Sillanpää. More contemporary authors and works include Hannele Mikaela Taivassalo's *The Five Knives of Andrei Krapl* (2007), Catharina Gripenberg's *Take My Hand, It Would be Strange* (2007), Ulrika Nielsen's *Between Linn Sand* (2006), Mikael Strömberg's *The Beautiful Cousins* (2008), Stefan Nyman's *Anna Is Online* (2008) and Monika Fagerholm's *Wonderful Women by the Water* (1997) and *The American Girl* (2010). The prolific nature of these writings and the creative undercurrent they inspire rank Finland as one of the most literate countries in the world.

Today, Finland is a *reading* society with various publications in the three ethnolinguistic categories—Finnish, Swedish and Sami. While the country sees "an annual publication of approximately 13,000-14,000 books a year, with over 4500 of them new works, there are more books published per capita in Finland than in any other country in the world except Iceland."[23] This may also explain the fact that since the founding of its first library in 1794 in Vaasa (the regional capital of Ostrobothnia in western Finland), the country has developed a comprehensive and well-funded network of public libraries carrying tens of millions of books and other materials.[24] The Finns are among the world's most pervasive users of library facilities. "Although they play an important role in public education, libraries also serve as cultural institutions and civic meeting centers for the public."[25] They also serve as centers of lifelong learning, especially for those who have already retired from the active world of work but continue to seek ways to avail themselves of a sense of awareness in emerging social and political issues.

10.2 Finnish Architecture and Design: Classicism with a Narrative

The Finns take great care in infusing their arts, designs, and sculptures with meaning and context. This quality is also very much present in their architectural and building designs and construction, in their clothing and apparel designs and in the materials that make these works possible. The development of modern Finnish architecture drew inspiration from the nationalist movement, and among its earliest pioneers were the internationally recognized Eliel Saarinen, whose works include the National Museum of Finland and the Helsinki Central Railway Station, the First Christ Church (Columbus, IN), Christ Church Lutheran (Minneapolis, MN) and Lars Sonck, whose architectural works include the St. Michael's Church (Turku), Tampere Cathedral (Tampere), Eira Hospital (Helsinki), Kallio Church (Helsinki), and Ainola (the residential home of Jean Sibelius and his wife Aino).

While notable architectural designers of the early twentieth century include the classicist Alvar Aalto (Lahti Church, 1950; Finlandia Hall, 1971), Erik Bryggman (Resurrection Chapel, Turku, 1941), and Johan Sigfrid Sirén, who built the Parliament Building in 1931, "there is a new generation of architects such as Juha Ilmari Leiviskä, known for his innovative churches, and the team of Pekka Helin and Tuomo Siitonen (built Nokia Headquarters, Espoo) whose flexible and adaptive working spaces are meant to encourage creative thinking."[26] There is also the imposing Tampere-talo (Tampere Hall—and home of the Tampere Philharmonic Orchestra) designed by architects Sakari Aartelo and Esa Piironen, which currently stands as the largest congress and concert hall in all of Scandinavia.

What has characterized Finnish architecture over the recent years is a deeply rooted deconstructive flair that can oftentimes be bold and provocative and at other times unassuming but ergonomically suited for the purpose for which it was designed. Drawing from a rich functionalist belief that form should follow function, the structure of the design would be expected to satisfy the purpose for which the project was meant to serve. Finnish designs, in apparel wear, furniture, architecture and machine tools, are considered living organisms oriented to serve specific purposes. The totality of the final product is seen as a reflection of the thought process, as well as the eye and hand coordination needed to transform an abstract ideal into a representative form. This has become a key of design thinking in contemporary de-urbanization projects, the development of *green cities* and *forest towns* and rational high-tech factories.

As in fashion clothing and apparel, functionalism permeates most Finnish designs and concepts in furniture, lightings, fixtures and decor. The willingness to try new things and to adapt to changing environmental contexts meant that a single design could incorporate various features and rhythms that could be considered eclectic, three-dimensional abstraction, parsimonious, eccentric or postmodern. With major influences from Sweden, Italy and Holland, Finnish designs encourage self-expression, simplicity and color neutrality. The emphasis is generally on natural or neutral colors, those that are not aggressively colorful and loud or seemingly suggestive of overt stereotypes or atypical ideological symbolisms. They embody a level of neo-expressionism without the alluring thickness of heavy impasto.

Nonetheless, if and when more robust colors are desired in an artistic presentation, it is often underplayed with a neutral colored background canvas that softens the aesthetic sharpness and tone. While neutral colors may seem to exude a sense of simplicity and innocence, they remain deceptively interpretive and functionalist in character. By projecting an aesthetic imagery of comfort and harmlessness, the neutrality of the colors holds the spectator in suspense, drawing him or her into a world far beyond the immediacy of the moment, yet a complex world that does not give up its true nature so easily. The point is that Finnish architecture, arts and designs, whether in industry, fashion, museums or other domestic and household applications, are imbued with meaning; and to engage them is to embark on an intriguing journey of discovery.

10.2.1 The Design Nation

It was not by a simple dint of good luck but by sheer effort and rare and creative imagination that Helsinki, the capital of Finland, was (along with surrounding cities of Espoo, Vantaa, Kauniainen and Lahti) named the 2102 Design Capital of the World. This recognition draws from a rich history and a trend-setting tradition that has seen Finnish designs as a much-cherished export commodity by connoisseurs, critics and the general public all over the world. Shedding its traditionally realist emphasis, the new Finnish trend embraces a modernist and postmodernist dimension that seeks to capture not only the physical world and its essential nature but also specific cultural moments that inform the society we live in. Like most Scandinavian art, Finnish design is a profound testimony reflecting nature in art, human imagination, the synchronization of form and function and obedience to tradition and cultural history. It has variously been characterized as embodying an insightful mix of abstractness, Slavic melancholy, romantic folklore, reductivism, minimalism and is playful, bold and quite idiosyncratic.[27]

Drawing from the general theme of the *Helsinki World Design Capital* 2012, it has become obvious that for the Finns, design goes beyond the individual, objects or fashion; it is also about the continuous quest to improve the conditions of cities as well as that of people who live in them. Hence, there has been a rising movement to expand the application of design from products to various systems and public services such as health care, public transportation, political systems, school systems and urban planning. The idea of '*embedding design in life*,' nonetheless, reinforces an essential recognition of the utility of design as a resource for social, economic and cultural development.[28] By infusing *body* and *soul* into inanimate objects in the attempt to reveal their inner qualities to the physical world, Finnish designs generally speak very loudly to the imagination even in the simplest forms. Using metaphor and allegory, they have often been used as symbolic statements or social commentary about the human condition and the state of the world.

10.2.2 Beyond Postmodernity?

The history of the country's design culture dates back to 1875 and showcases a heritage that is well documented at the Design Museum, the Arabia Museum in Helsinki (for ceramics) and the glass museum in Iittala and Nuutajärvi.[29] With the national emphasis on the training of designers and artists anchored at Aalto University, the Institute of Design in Lahti and also the University of Lapland's Faculty of Art and Design, "design has become a high-quality cultural export and represents an increasingly competitive factor for Finnish products and companies."[30] Today, many of the country's designers and artists are in great demand and work in various parts of the world. And to remain competitive, domestic companies such as textile manufacturer Marimekko allow their designers to be limited only by the extent of their individual imagination and to inject their own experiences and views of trends into the patterns.[31]

Finnish designers have become pacesetters, not only in ceramics and glass, but also in fashion, furniture, industrial designs and jewelry. Among these[32] are London-based Julia Lundsten, an internationally recognized shoe designer, Minna Parikka for shoes and handbags, Paola Suhonen (clothing and accessory), Hannu Kähönen (product design), and Sanna Annukka—all dominant figures in commercial art and graphic design; Timo Ripatti, Anti Kotilainen, Mikko Laakkonen, Terhi Tuominen, Eero Aarnio—in furniture design; the organic forms of Chao-Hsien Kuo and Eero Hinstanen (jewelry), and the hand-crafted silver touch of Tiina Arkko and Vesa Nilson (jewelry). In addition to popular Finnish design names such as Marimekko, Aalto and Arnio, one is also reminded of the creative works of Tapio Wirkkala (1915–1985), the father figure of Finnish applied art whose creative designs often reflect a sensitive dialogue between thought, hand, eye, and material.[33] He is credited with developing various crystal objects, glassware, vases and bottles with uniquely abstract artistic impressions.

There are also other designers of international repute such as Marja Metsovaara in textiles and Dora Ljung in *ryijy* (a type of knotted pile-weave rug).[34] Several of Finland's most successful designers have made their homes in foreign countries. Among these[35] are illustrator and graphic designer Lotta Nieminen, who resides in Brooklyn, NY; Barcelona-based Pietari Posti (illustrator and graphic design); London-based Klaus Haapaniemi (illustrator and graphic design); Paris-based Anna Ahonen and Katariina Lamberg; and Amsterdam-based Kustaa Saksi (commercial art and graphic design). And the most common denominator in the designs is that they all have the Finnish touch—expressive, emancipatory, yet open to varied interpretations.

Industrial design also plays a central role in Finnish lives and export trade. As a result, there have been continuous developments in materials science, synthetic wood products, polymers and building and construction materials—in an effort to conserve energy and resources as well as advance the course of environment management, recycling and sustainable development. Although "leading industrial design companies in Finland include Nokia, Kone, Metso, Fiskars, Iittala, Rocla, Suunto, and Vaisala; the greater Helsinki region alone boasts about 10,000 design professionals, and design-intensive companies of approximately 110,000–200,0000."[36] Whether the product is simple household furniture, a packaged food container, a glass bowl, a finger-nail clipper, a pencil sharpener, a belt buckle or a mobile phone handset, creative product design has certainly become an unintrusive mechanism that could be used to see into the soul of what it means to be Finnish.

10.3 The Sportsman as Hero: The Flying Finns

As in theatre, music, fashion, commercial arts and graphic design, Finland also has made its mark in national and international sports. Beyond traditional sports such as cross-country skiing, ice hockey, football, ski jumping, harness racing, water skiing, the national sports *pesäpallo* (a variant of American baseball) and fishing, the country has excelled in Olympic track and field—in which "the tradition of *Flying Finns* includes Hannes Kolehmainen, Ville Ritola, Lasse Viren and Paavo Johannes Nurmi—whose wining of nine gold and three silver medals in middle-and-long-distance running events at three Olympic games in the 1920s made him a national hero."[37] Nurmi's great achievements and dominance in world middle- and long-distance running at a time when Finland was still settling into the challenges of independence and statehood served as an embodiment of the country's aspirations vis-à-vis the international community. The country has honored him in several ways, including erecting a permanent statue of him in 1952 in front of the Olympic Stadium in Helsinki.

A century later, Finns have continued to make their mark on cross-country skiing following the Olympic exploits of athletes like cross-country skier Marja-Liisa Kirvesniemi (formerly Hämäläinen) who won seven Olympic medals in games held between 1984 and 1994, Matti Nykänen and Janne Ahonen in ski jumping and Pirkko Määttä who won two relay bronze medals (one each) at the 1984 Winter Olympic Games in Sarajevo, Yugoslavia, and the Winter Olympic Games held in Calgary, Alberta (Canada) in 1988. While ice hockey remains the most popular sports in the country, Finnish athletes have excelled and won medals in many other sports events including figure skating, snowboarding, wrestling and javelin throwing.

Over the years, Finnish motor sports rally and Formula One drivers have dominated and won championship medals at various international motor rally events. Among the most recognized names in world motor rally championships are Juha Kankkunen and Tommi Mäkinen; while Mika Häkkinen and Kimi Räikkönen won championships in Formula One racing. Besides winning championships, the idea of keeping active and fit through physical activity is a cherished cultural trait of the Finns because it is also a means of promoting good health and overall individual well-being. Increasing emphasis on physical activity as well as reducing obesity has always been a major focus of government policy, particularly of the Ministry of Social Affairs and Health and such non-profit organizations as the Tampere-based Urho Kekkonen Fitness Institute Foundation (UKK), Health SPA (Espoo) and several others.

10.3.1 The Sauna Is Ready

It is difficult to fully understand the aura surrounding the *sauna* in Finnish life until one has had the experience. In Finland, just like the air we breathe, the *sauna* is everywhere. It is state of mind that is unavoidably Finnish both in practice and in the logic surrounding its ubiquitous presence in the national culture and history. "Traditionally, the sauna was a sacred place for the Finns, used not only for the weekly sauna bath but also for ritual purposes as well as some domestic chores."[38] It is made mostly of birch or other sturdy wood beams. "Bathers sit on wooden benches, splashing water on the hot stones of the stove and whisking each other or themselves with birch branches, just as their ancestors would have done centuries earlier."[39] Each scoop of water splashed over the heated stones of the stove releases a moist and intensely heated hot vapor (*loyly*) and smoke throughout the room or enclosure. As your pores open up, sweat covers your whole body as your temperature starts to rise.[40] This is part of the therapeutic, or rather, purification experience generally associated with the sauna. "Saunas are usually sex segregated, but when done in a mixed sex scenario with non-family members, people will often wear a swimsuit or towel."[41] Either way, etiquette and discipline are expected to be strictly observed. And as the saying goes, in the sauna one must conduct oneself as if one were in a church.

Although in earlier times, the sauna was built as a separate structure away from the main building or farmhouse, today it is usually built as an integral part of residential, office or commercial buildings such as hotels, factories and apartment complexes. Unlike the original design, there are now modern versions of the sauna powered by electricity, oil or gas, and this has made it more convenient and less hazardous to be placed in more places than would ordinarily have been the case. While it is "believed that sauna bath provides healing for the mind and body, the experience is taken with an almost religious reverence and is a custom that is widespread among other Finnic peoples of the Baltic region—the Estonians, Karelians, Veps, and Livonians, as well as among Latvians and Lithuanians."[42]

Besides its social and cultural value, the sauna oftentimes offers moments of spiritual rejuvenation and quiet meditation that soothe the soul. Many also "believe that it helps to avoid disease and maintain good health by eliminating toxins through sweating,"[43] and "due to its social and relaxing effect, the sauna is also viewed as an excellent place for business negotiations and exchange of ideas, opinions, and consensus building."[44] Currently, "there are about 3.3 million saunas in Finland, and 99% of Finns undertake a sauna experience at least once a week and much more when they visit their summer cottage in the countryside."[45]

However, while the average Finnish guest is encouraged to try the sauna experience, it is generally considered a mark of respect and regard to acquiesce. As the author's own experience reveals while staying at the Hotel Cumulus in Helsinki on a Fulbright-related event, the heat from the sauna seemed as if it was drying up the oxygen in the air; my blood started to 'boil,' and I began to sweat profusely. I started to feel an impending imbalance or rather disequilibrium between my body temperature and that of the immediate surroundings. It is possible that although I was continuously splashing water onto the hot stone, I may have stoked the sauna heat too frequently, or I may have stayed a bit too long. Either way, as I left the sauna and headed for my room, I felt somewhat nauseous along the way such that I had to lean on the elevator wall for support. A quick and hearty drink of a bottle of cold water calmed my body temperature, and I soon fell asleep. There was no more sauna for me until I left Finland for the United States. And the message here is that, enjoy the sauna if and when you are privileged to do so, but be more reserved especially as a newcomer until you are better used to it.

10.3.2 Drinking Coffee with Cognac

The Finns are among the biggest consumers of coffee in the world, drinking an average of 12 kilograms per capita annually. While "coffee has persisted for more than three centuries as a favorite national drink, it is today seen as an integral part of Finnish culture and as a commodity that everyone is entitled to no matter the income level or changes in price."[46] Furthermore, the "fact that Finland is the only country in the world that has made coffee breaks at work statutory explains how strongly the people feel about the drink, and the center-stage role it plays in the most important events and festivities."[47] "Serving coffee is generally regarded as an act of hospitality; hence not offering it to a guest may be taken as an insult, especially among the older generations."[48] Although the country does not grow the coffee tree, it has nonetheless managed through these rituals to firmly embed it as a celebrated institution within the national cultural firmament.

It is not unusual or, perhaps, may even be customary for guests to be served a cup of coffee along with a glass of cognac at Finnish events. I have pondered several times over whether there is any cultural significance of drinking coffee and cognac at the same time? I know that coffee is generally associated with caffeine which is a stimulant (a kind of 'picker upper'), but cognac as most other alcoholic beverages is a relaxant and has been associated with "improving the vasodilatory reactivity of the arteries."[49] While there is no basis for seeking further physiological explanation in this regard, the issue is on how one should interpret or make sense of their combined use knowing that they have contrasting effects on the physical agility of the human body. Nonetheless, and if one can be permitted to speculate, it is possible that the cultural implication is that it serves as a way of relaxing the mind and of getting one more focused on the event and the social mood of the moment. Alternatively and even if there is no cultural relevance, it should be seen as one more intriguing phenomenon about this land of great surprises.

10.3.3 To Estonia and Back

The Finnish lifestyle has been evolving over the years in fundamental ways. Much of this has been complicated by the rising level of drinking and smoking across various demographic and social categories in the country. In a typical gathering or just walking along a crowded street, it would not be unusual to discover, albeit to the dismay of a non-smoker, that you are surrounded by the odor from cigarette smoke; hence, the discomfort of passive smoking remains most evident. The Ministry of Social Affairs and Health is not oblivious to this development and to the fact that alcoholism, excessive smoking and obesity are major health and behavioral problems in Finland. These are long-term issues with which the welfare state must contend now and perhaps for the foreseeable future. Compared to neighboring Estonia just south of the border, the cost of alcohol is relatively cheaper than in Finland, particularly in the Greater Helsinki region. As a result, a de facto informal economy (with official cognizance) has developed between the two countries where scores of Finns make the two hour trip across the Gulf of Finland by ferry or cruise ship to Tallin (capital of Estonia) to load up on all manners of beer, liquor, and cigarettes and bring these back to Finland. Since they are assumed to be for "personal" consumption, they do not generally pay customs tariffs or taxes for most of what they bring in.

This informal economy seems to have comfortably embedded itself into the formal political and bilateral relationship between the two countries such that authorities in Helsinki have been reluctant to address the issue in a more forceful way. They are aware that the increased access to alcohol and cigarettes from Estonia accentuates the associated health problems in Finland and puts inordinate pressure on costs as well as the health care infrastructure. It also undermines existing policies meant to reduce alcoholism and addiction among the Finnish population at great expense to the welfare system. Nonetheless, other considerations have served to politicize this informal trade relation that could make it much harder to limit people's access to cheap alcohol and cigarettes from Estonia.

First, being one of the new post-Soviet independent Baltic States (along with Latvia and Lithuania), Estonia needs all the economic boost it can get, especially from her wealthier and more advanced neighbor to the north. The Finnish government could impose a highly restrictive quota or tariff on the alcohol transit from Estonia to Finland, but there is little political will to do so. Furthermore, the whole enterprise comes under the protective umbrella of the two governments' policies and promotion of interstate tourism. Various private sector ferry and cruise line companies such as Tallink Silja, Viking Line and the Eckerö Line make a combined 12 sailings per day on the Helsinki to Tallin route and would lobby against any regulation that would restrict transportation or trade across the Gulf. Second, Finnish policy makers see Estonia as the tip of the spear for a regionally based economic community or free trade area with Helsinki as the epicenter. Third, the Estonians are, like the Finns, part of the larger Finnic ethno-cultural lineage and could be the benefactors of a more liberal *be-thy-brother's keeper* sentiment in Finland. These considerations may have collectively trumped a more aggressive Finnish stance on this issue.

10.4 A Concluding Postscript

In the end, it can still be asked: Who are the Finns, what do they want, and what is it that drives them? It is somewhat of an over simplification when they are described as being insular, taciturn, friendly and generally not given to 'small-talk.' The Finns are probably a little bit of this and a little bit of that, but are more complex and difficult to pin down to a specific individual or cultural attribute but are generally driven by two essential factors: pride of country and a keenness to educate the visitor about how their history and experience has shaped their national life. The average Finn may also be overtly drawn to a certain kind of introversion that is existentialist and self-referential. This may be what most people cavalierly attribute to their being inwardly and reserved. For example, if a person's character or attitude is spread on a continuum linked by two extremes, I would rank the average Finn's disposition in the middle, shifting either left or right, but in accordance with the subject matter of interest and the realities surrounding it. I would say that the average Finn's attitude hovers at the mid-point where an individual's natural skepticism meets but does not overtake the practical reality of the moment. It is the not-so-obvious but lingering skepticism that is at the core of their evident inquisitiveness and curiosity to constantly seek knowledge and understanding of nature as well as its relationship to the mechanical world we live in. Trust, including friendship, is not something they give away so easily; these take time to build and must be earned and continuously validated by remaining true to the initial premise or conditions under which they were made.

On the question of being insular or taciturn, this is probably one of the most exaggerated characteristics of the Finns. Can they *talk*? Yes, you only need to initiate or provoke a conversation, and they will certainly engage in the discussion. The many Finns with whom I have had the great occasion of discussing and deliberating on a series of issues were quite engaged in the conversation to the extent that I can say, more often than not, that I was put in a position of being a listener more than a participant. But they do not talk just for the sake of talking in terms of what is generally known as 'small talk.' When they talk, they are reflective about what they say and at the same time sensitive regarding how you may react or make sense of what they say. They expend a lot of mental energy because they want to make sure that each and every word they say carries the power and weight of their conviction and understanding on the subject matter. They are not generally given to satire, but are serious engagers in the art of discussion, focused but willing to see it to the end. While it may be difficult to state whether this attribute is a form of virtue or weakness, what is undeniable is that maintaining eye contact and engaging each discussion as if it was a presentation is an essential part of the average Finn's social and deliberative character. To the extent that being reserved in certain respects is not the same as being habitually shy, the idea that Finns are essentially shy may thus need further empirical validation to be accepted as an absolute.

What about keeping time and punctuality? The average Finn is likely to keep strictly to the time of an appointment. Guests are expected to do the same. As is customary in most cases, you will be served coffee, water and one of the many varieties of Finnish flatbread or some other snack like potato pasties (*perunapiirakka*) as refreshment. As you get down to the discussion, they have a tendency to stick to the basics while at the same time maintaining eye contact, just in the same way that Americans are generally known to do during conversations. But while this characteristic is not any different from most other societies or cultures, I believe that the general attitude to keeping with the time of appointment is taken more seriously in the United States and, in fact, more severely enforced and consequential than in Finland. In the US, if you are late or miss an important appointment without a convincingly credible excuse, you may not have another chance to reschedule, no matter how important the issue. Hence, I do not see the Finns as having a moratorium on punctuality as virtue. They emphasize it, but so do most other cultures.

There is another aspect of Finnish life that is oftentimes muted both in its historical relevance and for its latent implication regarding Finland's foreign policy and national security interest. Even though there is relative peaceful coexistence between Finland and its neighbor to the East (Russia), it remains well prepared to provide for its own security. The central strategy is not to fight but to do everything possible to avoid war. On the other hand, and should the country be attacked, it is prepared to defend itself and to make it a very expensive proposition for any would-be aggressor. In preparation, there is a requirement for all able-bodied males to serve in the armed forces at some point in their adolescent lives. Traditionally, Finnish soldiers are trained in forest warfare and survival tactics in harsh climatic conditions, but because of the changing nature of war and especially the proximity of its capital (Helsinki) to Russia, the Finns have assiduously adopted the model of urban warfare as part of their tactical doctrine.

While some soldiers are trained in forest warfare, others receive reserve or specialist training that allows them to learn to defend major urban centers such as Helsinki or Turku. Because the vital institutions of government are in Helsinki, a series of redundancies have been built into the command and control system, and defensive techniques of urban warfare have become dominant. Men and women who may currently be working in civilian institutions are always on call and have their military uniforms and outfits with them in their homes. They already know where key weapon systems are and what to do and where to go as soon as they are called up for any emergency situation. The Finns always anticipate the "what if?" but then they prepare for it as a means of containing the uncertainty and the contingencies that the unknown may bring. Theirs is a forward-thinking mentality.

Notes

1 The first sentence of the Finnish national anthem reads thus: "Oh our land, Finland, land of our birth, ring out the golden word!" The lyrics were written by the Finnish poet Johan Ludvig Runeberg in 1848, while the corresponding music was by the German-born composer Fredrik Pacius.

2 Ruohonen, A., S. Salzano, and P. Marten. 2014. Bring on the Finnish summer festivals, *This is Finland*, pp. 1–6. http://finland.fi/Public/default.aspx?contentid=307471&nodeid=44497&culture=en-US. Accessed on August 12, 2015.

3 Lönnrot, E. 2008. *The Kalevala* (Translated by Keith Bosley). New York, NY: Oxford University Press, p. 469.
4 http://matadornetwork.com/life/x-things-not-know-finland. Accessed on August 16, 2015.
5 http://www.fime.fi/en/culture/culture-in-finland. Accessed on August 12, 2015.
6 Ibid.
7 Ibid.
8 http://www.everyculture.com/Cr-GA/Finland.html. Accessed on August 12, 2015, p. 3.
9 Ibid.
10 http://matadornetwork.com/life/x-things-not-know-finland. Accessed on August 16, 2015.
11 Ibid.
12 http://www.everyculture.com/Cr-GA/Finland.html. Accessed on August 12, 2015, p. 8.
13 Ibid.
14 http://www.britannica.com/print/article/207424. Accessed on August 12, 2015, p. 4.
15 Amberla, K. 1999. Facts about Finnish music life. *Finnish Music Quarterly,* 3, September, p. 1. http://fmq.fi/1999/09/facts-about-finnish-music-life.
16 http://www.britannica.com/print/article/207424, p. 6. Accessed on August 12, 2015.
17 Ibid. p. 5.
18 YouTube: https://www.youtube.com/watch?v=5pQMW8UQYZw&feature=youtu.be.
19 http://finland.fi/public/default.aspx?contentid=160075&contentlan=2&culture=en-US, p. 2. Accessed on August 12, 2015.
20 Ibid.
21 http://www.britannica.com/biography/Aleksis-Kivi, p. 1. Accessed on August 12, 2015.
22 http://www.everyculture.com/Cr-GA/Finland.html, p. 7. Accessed on August 12, 2015.
23 http://finland.fi/Public/default.aspx?contentid=16007&nodeid=44495&culture=en-US, p. 1. Accessed on August 12, 2015.
24 http://www.britannica.com/print/article/207424, p. 6. Accessed on August 12, 2015.
25 Ibid.
26 Ibid.
27 Marten, P. (ed.) 2012. *Welcome to Finland: The design nation*. Helsinki, Finland: The Ministry of Foreign Affairs.
28 http://finland.fi/Public/default.aspx?contentid=209426&nodeid=44495&culture=en-US, p. 2. Accessed on August 12, 2015.
29 Marten, P. (ed.) 2012. *Welcome to Finland: The design nation.*
30 http://finland.fi/Public/default.aspx?contentid=209426&nodeid=44495&culture=en-US. Accessed on August 12, 2015, p. 2.
31 Ibid.
32 Marten, P. (ed.) 2012. *Welcome to Finland: The design nation.*
33 http://finland.fi/Public/default.aspx?contentid=16007&nodeid=44495&culture=en-US, p. 2. Accessed on August 12, 2015.
34 http://www.britannica.com/print/article/207424, p. 6. Accessed on August 12, 2015.
35 http://finland.fi/Public/default.aspx?contentid=301809&nodeid=44495&culture=en-US, p. 2. Accessed on August 12, 2015.
36 Marten, P. (ed.) 2012. *Welcome to Finland: The design nation.*
37 http://www.britannica.com/print/article/207424, p. 7. Accessed on August 12, 2015.
38 Ibid.
39 Ibid.
40 Bosworth, M. 2013. Why Finland loves saunas. *Magazine*, October 1, p. 1.
41 http://independenttravelcats.com/2015/03/22/visiing-finland-in-winter-top-winter-activities-in-finland. Accessed on August 18, 2015.
42 http://www.britannica.com/print/article/207424, p. 4. Accessed on August 12, 2015.

43 http://pureinsideout.com/sauna-finnish.html. Accessed on August 16, 2015.
44 Ibid.
45 Bosworth, M. Why Finland loves saunas, p. 2.
46 Ojaniemi, T. 2010. Coffee as a Finnish institution. *Finnish Institutions research paper*, FAST Area Studies Program, Department of Translation Studies, University of Tampere, 1–7, p. 1. http://www15.uta.fi/FAST/FIN/GEN/to-coffee.html. Accessed on July 31, 2015.
47 Ojaniemi, T. Coffee as a Finnish institution, p. 1. http://www15.uta.fi/FAST/FIN/GEN/to-coffee.html. Accessed on July 31, 2015.
48 Ibid. p. 4.
49 Kiviniemi, T. O. et al. 2008. Effects of cognac on coronary flow reserve and plasma antioxidant status in healthy young men. *Cardiovascular Ultrasound*, 6: 25. http://www.ncbi.nlm.nih.gov/pmc/articles/PMC2427009. Accessed on July 31, 2015.

Appendix A Chronological History of Presidents and Prime Ministers of Finland since Independence

Presidents	*Tenure of Office*
Kaarlo Ståhlberg	July 27, 1919 – March 1, 1925
Lauri Relander	March 1, 1925 – March 1, 1931
Pehr Evind Svinhufvud	March 1, 1931 – March 1, 1937
Kyösti Kallio	March 1, 1937 – December 19, 1940
Risto Ryti	December 19, 1940 – August 4, 1944
Carl Gustaf Mannerheim	August 4, 1944 – March 11, 1946
Juho Paasikivi	March 11, 1946 – March 1, 1956
Urho Kekkonen	March 1, 1956 – January 27, 1982
Mauno Koivisto	January 27, 1982 – March 1, 1994
Martti Ahtisaari	March 1, 1994 – March 1, 2000
Tarja Halonen	March 1, 2000 – March 1, 2012
Sauli Niinisto	March 1, 2012 –
Prime Ministers	
Pehr Evind Svinhufvud	November 27, 1917 – May 27, 1918
Juho Kusti Paasikivi	May 27, 1918 – November 26, 1918
Lauri Ingman	November 26, 1918 – April 17, 1919
Kaarlo Castrén	April 17, 1919 – August 15, 1919
Juho Vennola	August 15, 1919 – March 15, 1920
Rafael Erich	March 15, 1920 – April 9, 1921
Juho Heikki Vennola	April 9, 1921 – June 2, 1922
Kaarlo Cajander	June 2, 1922 – November 14, 1922
Kyösti Kallio	November 14, 1922 – January 18, 1924
Kaarlo Cajander	January 18, 1924 – May 31, 1924
Lauri Ingman	May 31, 1924 – March 31, 1925
Antti Tulenheimo	March 31, 1925 – December 31, 1925
Kyösti Kallio	December 31, 1925 – December 13, 1926
Väinö Tanner	December 13, 1926 – December 17, 1927
Juho Emil Sunila	December 17, 1927 – December 22, 1928
Oskari Mantere	December 22, 1928 – August 16, 1929
Kyösti Kallio	August 16, 1929 – July 6, 1930
Pehr Evind Svinhufvud	July 6, 1930 – February 16, 1931
Juho Heikki Vennola	February 16, 1931 – March 21, 1931
Juho Sunila	March 21, 1931 – December 14, 1932
Toivo Mikael Kivimäki	December 14, 1932 – October 7, 1936

Kyösti Kallio	October 7, 1936 – March 12, 1937
Kaarlo Cajander	March 12, 1937 – December 1, 1939
Risto Ry`ti	December 1, 1939 – December 21, 1940
Karl Walden	December 21, 1940 – January 3, 1941
Johan Rangell	January 3, 1941 – March 5, 1943
Edwin Linkomies	March 5, 1943 – August 8, 1944
Verner Hackzell	August 8, 1944 – September 21, 1944
Jonas Castrén	September 21, 1944 – November 17, 1944
Juho Kusti Paasikivi	November 17, 1944 – March 26, 1946
Mauno Pekkala	March 26, 1946 – July 29, 1948
Karl August Fagerholm	July 29, 1948 – March 17, 1950
Urho Kekkonen	March 17, 1950 – November 17, 1953
Sakari Tuomioja	November 17, 1953 – May 5, 1954
Gustaf Törngren	May 5, 1954 – October 20, 1954
Urho Kekkonen	October 20, 1954 – February 17, 1956
Karl Fagerholm	February 17, 1956 – May 27, 1957
Väinö Sukselainen	May 27, 1957 – November 29, 1957
Rainer von Fieandt	November 29, 1957 – April 26, 1958
Reino Kuuskoski	April 26, 1958 – August 29, 1958
Karl August Fagerholm	August 29, 1958 – January 13, 1959
Väinö Sukselainen	January 13, 1959 – May 19, 1961
Vihtori Luukka	May 19, 1961 – July 14, 1961
Martti Miettunen	July 14, 1961 – April 13, 1962
Ahti Karjalainen	April 13, 1962 – December 18, 1963
Reino Ragnar Lehto	December 18, 1963 – September 12, 1964
Johannes Virolainen	September 12, 1964 – May 27, 1966
Kustaa Paasio	May 27, 1966 – March 22, 1968
Mauno Koivisto	March 22, 1968 – May 14, 1970
Teuvo Aura	May 14, 1970 – July 15, 1970
Ahti Karjalainen	July 15, 1970 – October 29, 1971
Teuvo Aura	October 29, 1971 – February 23, 1972
Kustaa Paasio	February 23, 1972 – September 4, 1972
Kalevi Sorsa	September 4, 1972 – June 13, 1975
Keijo Liinamaa	June 13, 1975 – November 30, 1975
Martti Miettunen	November 30, 1975 – May 15, 1977
Kalevi Sorsa	May 15, 1977 – May 26, 1979
Mauno Koivisto	May 26, 1979 – January 27, 1982
Eino Uusitalo	January 27, 1982 – February 19, 1982
Kalevi Sorsa	February 19, 1982 – April 30, 1987
Harri Holkeri	April 30, 1987 – April 26, 1991
Esko Aho	April 26, 1991 – April 13, 1995
Paavo Lipponen	April 13, 1995 – April 17, 2003
Anneli Jäätteenmäki	April 17, 2003 – June 24, 2003
Matti Vanhanen	June 24, 2003 – June 22, 2010
Mari Kiviniemi	June 22, 2010 – June 22, 2011
Jyrki Katainen	June 22, 2011 – June 24, 2014
Alexander Stubb	June 24, 2014 – May 29, 2015
Juha Sipila	May 29, 2015 –

Source: Compiled by author based on information from The Countries Wiki Navigation, publications of the Republic of Finland, and the Parliament.

Appendix B
English Version
Survey/Questionnaire

The Interaction of Technology and Culture in Shaping Civic Attitudes toward Political Institutions and Policy Outcomes: Lessons from the Finnish Experience

Case No: ____________ **Please circle your choice.**

1. At what level of government/non-government do you work?
   1. Federal 2. State
   3. Local Government 4. Private Sector
   5. Non-Profit 6. Self-Employed __________
   7. Other (Specifiy) __________

2. What is your job title?
   1. Director 2. Manager 3. Supervisor
   4. Administrator 5. Other (Specify) __________

3. What is your gender?
   1. Male 2. Female

4. What is your <u>region</u> of origin? __________

5. Please write below the name of region/city/town of <u>residence</u>:
   1. Region __________ 2. City/Town __________

6. Do you live in an urban or rural area?
   1. Urban 2. Rural

7. What is the dominant form of productive activity in your place of residence?
   1. Industry 2. Agriculture 3. Business/Commerce
   4. Government/Administration 5. Fishing

8. What is your highest education level completed?

High School Diploma	Bachelor's Degree	Master's Degree	Doctorate
1	2	3	4
Technical/Other			
5			

9. How many years of experience do you have in your current position?

0–2	3–5	6–10	11–15	More than 16 years
1	2	3	4	5

10. What political party do you belong to, if any?
    1. Center Party
    2. Christian Democrat
    3. Green League
    4. Left Alliance
    5. National Coalition Party
    6. Social Democratic Party
    7. Swedish People's Party
    8. The Finns Party
    9. Independent/Non-Partisan

11. Do you consider yourself a party activist?
    1. Yes
    2. No

12. Which of these represents your views on issues of politics and governance?
    1. Conservative
    2. Liberal
    3. Socialist
    4. Egalitarian
    5. Other (specify) _______

13. Which of these represents your views on the economy and economic policy?
    1. Conservative
    2. Liberal
    3. Socialist
    4. Egalitarian
    5. Others (specify) _______

14. Which of these represents your ethnicity?
    1. Finn
    2. Swede
    3. Russian
    4. Estonian
    5. Other (Specify) _______

[**Please continue the survey on the following pages**]:

Social Capital:

SC 15. How much of the time do you think you can trust people (fellow citizens) in general?

1 Just about always
2 Most of the time
3 Some of the time
4 Hardly ever
5 Don't know

SC 16. How much of the time do you think you can trust the national government to do what is right?

1 Just about always
2 Most of the time
3 Some of the time
4 Hardly ever
5 Don't know

SC 17. How much of the time do you think you can trust your local/regional government to do what is right?

1 Just about always
2 Most of the time
3 Some of the time
4 Hardly ever
5 Don't know

SC 18. How much trust do you have in government agencies' ability to serve the people in general?

1 Trust them a lot
2 Trust them sometimes
3 Trust them only a little
4 Don't trust them at all
5 Don't know

SC 19. How about the NATIONAL government? How much of the time do you think you can trust the NATIONAL government to do what is right?

1 Just about always
2 Most of the time
3 Some of the time
4 Hardly ever
5 Don't know

SC 20. I feel comfortable with my co-workers.

1 Strongly agree
2 Agree
3 Indifferent
4 Disagree
5 Strongly disagree
6 Don't know

SC 21. I feel comfortable with my fellow citizens and clients.

1 Strongly agree
2 Agree
3 Indifferent
4 Disagree
5 Strongly disagree
6 Don't know

SC 22. I sense that others in the COUNTRY/community are comfortable with my agency's or organization's services.

1 Strongly agree
2 Agree
3 Indifferent

4 Disagree
5 Strongly disagree
6 Don't know

SC 23. I sense that my clients/citizens are comfortable with my agency's or organization's services.
1 Strongly agree
2 Agree
3 Indifferent
4 Disagree
5 Strongly disagree
6 Don't know

SC 24. How much do you think that your community/fellow citizens share the same values, identifications and world views as you (personally)?
1 A lot
2 Sometimes
3 A little
4 Not at all
5 Don't know

SC 25. How much do you think that your community/fellow citizens share the same basic values, identifications and world views as reflected in your agency or organizational mission?
1 A lot
2 Sometimes
3 A little
4 Not at all
5 Don't know

SC 26. How much do you think that your community/fellow citizens share the same basic values and identifications as you in terms of what is happening in the country today AND how to address most social problems?
1 A lot
2 Sometimes
3 A little
4 Not at all
5 Don't know

SC 27. How *confident* are you that most people understand what is happening in the country today and what the future holds for its citizens?
1 A lot
2 Sometimes
3 A little
4 Not at all
5 Highly uncertain
6 Don't know

Technology/ICT:

T 28. In 2011, the percentage of individuals in Finland using the Internet is about 89.37%. How often do you use the Internet to source information about the government or any of its agencies?
1. Always 2. Sometimes 3. Infrequently
4. A few occasions 5. Not at all

T 29. How often do you use the Internet to access information about the Ministry of Education?
1. Always 2 Sometimes 3 Infrequently
4. A few occasions 5. Not at all

T 30. How often do you use the Internet to access information about the Ministry of Health?
1. Always 2. Sometimes 3. Infrequently
4. A few occasions 5. Not at all

T 31. What is your most common means of receiving information about national policy?
1. Newspapers 2. Internet 3. Online media
4. Cell phones 5. E-mail 6. Other (specify) _______

T 32. Which of these media do you most commonly use in your daily or regular communications?
1. E-mail 2. Cell Phone 3. Texting
4. Skype 5. Face to Face 6. Other (Specify) _______

T 33. Through which of these media are you most comfortable in communicating with your fellow co-workers and/or citizens?
1. E-mail 2. Cell Phone 3. Texting
4. Skype 5. Face to Face 6. Other (Specify) _______

T 34. Would you agree or disagree that using e-mail, cell phone, texting, Skype and other electronic media in communications makes Finnish government and society more functional?
1 Strongly agree
2 Agree
3 Indifferent
4 Disagree
5 Strongly disagree
6 Don't know

T 35. About 85% of Finnish households and individuals have access to and use computers. What do you mostly use your computer for?
1. Work 2. Online Purchasing 3. Banking
4. Bill Payment 5. Communicate with People 6. All of the Above

Institutional Effectiveness/Outcome:

IE/O 36. How often do you use the Internet to evaluate the performance or effectiveness of education policy in Finland?

1. Always 2. Sometimes 3. Infrequently
4. A few occasions 5. Not at all

IE/O 37. In regard to IE/O 35 (above) would you say that your evaluation has generally been?

1. Positive 2. Moderately Positive 3. Average
4. Moderately Negative 5. Negative

IE/O 38. On a scale of 1–5 (**5 being the highest**), how would you rank the effectiveness of education policy in Finland?

1. Poor 2. Satisfactory 3. Average
4. Excellent 5. Superior

IE/O 39. How often do you use the Internet to evaluate the performance or effectiveness of health policy in Finland?

1. Always 2. Sometimes 3. Infrequently
4. A few occasions 5. Not at all

IE/O 40. In regard to IE/O 38 (above) would you say that your evaluation has generally been?

1. Positive 2. Moderately Positive 3. Average
4. Moderately Negative 5. Negative

IE/O 41. On a scale of 1-5 (**5 being the highest**), how would you rank the effectiveness of health policy in Finland?

1. Poor 2. Satisfactory 3. Average
4. Excellent 5. Superior

Policy Dissonance:

PD 42. The information I receive from the national government about national issues is consistent and easily understood.

1 Strongly agree
2 Agree
3 Indifferent
4 Disagree
5 Strongly disagree
6 Don't know

PD 43. The information I receive from the local government about local government issues is consistent and easily understood.

1 Strongly agree
2 Agree
3 Indifferent

4 Disagree
5 Strongly disagree
6 Don't know

PD 44. I believe that lack of common processes and terminology used between the national and regional governments is currently an obstacle to achieving effective administrative coordination.
1 Strongly agree
2 Agree
3 Indifferent
4 Disagree
5 Strongly disagree
6 Don't know

PD 45. In your experience, how often do local conditions dictate that implementation of national policies be different from specific guidelines given by national agencies/officials?
1 Just about always
2 Most of the time
3 Some of the time
4 Hardly ever
5 Don't know

PD 46. What do you consider to be the greatest current or potential threat to the country at the moment?

1 Economic Depression	2 Unemployment	3 Security/Terrorism
4 Corruption	5 Inflation	6 Demographic Changes

PD 47. In terms of infrastructural development, which of these should receive the greatest priority from the **national** government?
1 Water Supplies
2 Electricity
3 Reconstruction of Educational Institutions
4 Restoring Public Service/Government Accountability
5 Provision of Good Health Care
6 Helping/Strengthening the Technology Industry
7 Rebuilding Road/Transportation Network

PD 48. In terms of infrastructural development, which of these should receive the greatest priority from the **regional** governments?
1 Water Supplies
2 Electricity
3 Reconstruction of Educational Institutions
4 Restoring Public Service/Government Accountability
5 Provision of Good Health Care
6 Helping/Strengthening the Technology Industry
7 Rebuilding Road/Transportation Network

Epistemological Relativism:

ER 49. In your opinion, what do you think is the foremost problem in the world?

Global Hunger	Globalization	Climate Change
1	2	3
Transborder Immigration	War and Conflict	Drug/Human Trafficking
4	5	6

ER 50. In your opinion, what do you think is the foremost problem in Finland?

Technology	Globalization	Climate Change
1	2	3
Transborder Immigration	Culture Change	Drug/Human Trafficking
4	5	6

ER 51. At this point, do you believe we have an accurate understanding of the multiple problems that confront the human society?

Strongly Disagree	Disagree	Neither Agree/Disagree
1	2	3
Agree	Strongly Agree	
4	5	

ER 52. If we can understand some of the problems that face the world, do you believe we can have a unified/common answer to solve them?

Strongly Disagree	Disagree	Neither Agree/Disagree
1	2	3
Agree	Strongly Agree	
4	5	

ER 53. Do you believe that the traditional culture and norms of Finland should be preserved even in the face of increasing technology and globalization?

Strongly Disagree	Disagree	Neither Agree/Disagree
1	2	3
Agree	Strongly Agree	
4	5	

ER 54. Do you believe that we should use technology to help Finnish society to adapt for the future?

Strongly Disagree	Disagree	Neither Agree/Disagree
1	2	3
Agree	Strongly Agree	
4	5	

ER 55. Do you believe that answers or remedies to various social problems are yet to be found and we should continue to look for them?

Strongly Disagree	Disagree	Neither Agree/Disagree
1	2	3
Agree	Strongly Agree	
4	5	

ER 56. The truth about the world we live in has already been discovered.

Strongly Disagree	Disagree	Neither Agree/Disagree
1	2	3
Agree	Strongly Agree	
4	5	

ER 57. Technology can help mankind acquire a better understanding of the world we live in.

Strongly Disagree	Disagree	Neither Agree/Disagree
1	2	3
Agree	Strongly Agree	
4	5	

ER 58. Technology can make the world we live in more complex hence more difficult to understand and manage.

Strongly Disagree	Disagree	Neither Agree/Disagree
1	2	3
Agree	Strongly Agree	
4	5	

ER 59. The truth about the world we live in is what it is and neither technology nor mankind can fully understand it.

Strongly Disagree	Disagree	Neither Agree/Disagree
1	2	3
Agree	Strongly Agree	
4	5	

ER 60. The values and traditions of other ethnic groups are just as good and acceptable as those of my own ethnic group.

Strongly Disagree	Disagree	Neither Agree/Disagree
1	2	3
Agree	Strongly Agree	
4	5	

ER 61. The potential benefits of technology to human society far outweigh its potential disadvantages.

Strongly Disagree	Disagree	Neither Agree/Disagree
1	2	3
Agree	Strongly Agree	
4	5	

Additional Comments: [Please feel free to provide additional comments that you may have]

Appendix C
Finnish Version
Liite C: Kyselylomake

"The Interaction of Technology and Culture in Shaping Civic Attitudes towards Political Institutions and Policy Outcomes: Lessons from the Finnish Experience"

Tapausnumero:____________

Ympyröi valitsemasi numero vastausvaihtoehdoista.

1. Millä hallinnon tasolla/minkälaisessa organisaatiossa työskentelet?
   1. Liittovaltio (ei eikoske Suomea) 2. Valtio
   3. Paikallishallinto 4. Yksityissektori 5. Järjestö, yhdistys
   6. Yrittäjä 7. Muu, mikä?

2. Mikä on asemasi?
   1. Johtaja 2. Päällikkö 3. Esimies
   4. Hallinnoija 5. Muu, mikä?

3. Sukupuoli
   1. Mies 2. Nainen

4. Mistä maakunnasta olet kotoisin? ________________________

5. Mikä on asuinpaikkakuntasi? Kirjoita vastaus alle.
   1. Maakunta 2. Kaupunki

6. Onko asuinpaikkasi kaupunkialuetta vai maaseutua?
   1. Kaupunkialuetta 2. Maaseutua

7. Mikä on asuinalueesi pääasiallinen toimiala?
   1. Teollisuus 2. Maanviljely 3. Kauppa tai liiketoiminta
   4. Julkishallinto / hallinto 5. Kalastus

8. Mikä on ylin suorittamasi tutkinto?
   1. Ylioppilas 2. Alempi korkeakoulututkinto
   3. Ylempi korkeakoulututkinto 4. Tohtori
   5. Ammattikoulu / muu

9. Kuinka kauan olet työskennellyt tämänhetkisessä tehtävässäsi?
   1. 0–2 vuotta 2. Kristillisdemokraatit
   3. 6–10 vuotta 4. 11–15 vuotta 5. Yli 16 vuotta

10. Oletko jonkin puolueen jäsen?
   1. Keskusta 2. Kristillisdemokraatit
   3. Vihreät 4. Vasemmistoliitto
   5. Kokoomus 6. SDP
   7. RKP 8. Perussuomalaiset
   9. Riippumaton / en kuulu mihinkään poliittiseen puolueeseen

11. Pidätkö itseäsi puolueaktiivina?
   1. Kyllä 2. En

12. Mikä seuraavista vastaa näkemyksiäsi politiikasta ja päätöksenteosta?
   1. Konservatiivinen 2. Liberaali 3. Sosialistinen
   4. Egalitaarinen 5. Muu, mikä?

13. Mikä seuraavista vastaa parhaiten näkemystäsi taloudesta ja talouspolitiikasta?
   1. Konservatiivinen 2. Liberaali 3. Sosialistinen
   4. Egalitaarinen 5. Muu,mikä?

14. Mikä seuraavista vastaa kansallisuuttasi?
   1. suomalainen 2. Ruotsalainen 3. Venäläinen
   4. virolainen 5. Muu, mikä?

[Kysymykset jatkuvat seuraavilla sivuilla.]

Sosiaalinen pääoma

SC 15. Kuinka usein olet sitä mieltä, että voit yleisesti ottaen luottaa muihin ihmisiin (oman maasi kansalaisiin)?
1. Melkein aina
2. Useimmiten
3. Toisinaan
4. Harvoin
5. En osaa sanoa

SC 16. Kuinka usein voit mielestäsi luottaa siihen, että maasi hallitus toimii oikein?
1. Melkein aina
2. Useimmiten
3. Toisinaan
4. Harvoin
5. En osaa sanoa

SC 17. Kuinka usein voit mielestäsi luottaa siihen, että asuinpaikkasi paikallishallinto toimii oikein?
1. Melkein aina
2. Useimmiten
3. Toisinaan
4. Harvoin
5. En osaa sanoa

SC 18. Kuinka paljon luotat yleisesti ottaen valtiollisten toimijoiden kykyyn palvella ihmisiä?
1. Paljon
2. Joskus
3. Vähän
4. En lainkaan
5. En osaa sanoa

SC 19. Entä paikallishallinnon tasolla? Kuinka usein voit mielestäsi luottaa siihen, että asuinpaikkasi paikallishallinto toimii oikein?
1. Melkein aina
2. Useimmiten
3. Toisinaan
4. Harvoin
5. En osaa sanoa

SC 20. Luotan työtovereihini.
1. Vahvasti samaa mieltä
2. Samaa mieltä
3. En samaa enkä erimieltä
4. Eri mieltä
5. Vahvasti eri mieltä
6. En osaa sanoa

SC 21. Luotan kanssaihmisiin ja asiakkaisiin.
1. Vahvasti samaa mieltä
2. Samaa mieltä
3. En samaa enkä erimieltä
4. Eri mieltä
5. Vahvasti eri mieltä
6. En osaa sanoa

SC 22. Uskon, että muut samassa MAASSA tai yhteisössä asuvat luottavat organisaationi palveluihin.
1. Vahvasti samaa mieltä
2. Samaa mieltä
3. En samaa enkä erimieltä

4. Eri mieltä
5. Vahvasti eri mieltä
6. En osaa sanoa

SC 23. Uskon, että asiakkaat / muut kansalaiset luottavat organisaationi palveluihin.
1. Vahvasti samaa mieltä
2. Samaa mieltä
3. En samaa enkä erimieltä
4. Eri mieltä
5. Vahvasti eri mieltä
6. En osaa sanoa

SC 24. Missä määrin arvelet, että muilla yhteisösi jäsenillä tai muilla kansalaisilla on samat arvot, samaistumisen kohteet ja maailmankuva kuin sinulla?
1. Paljon
2. Joskus
3. Vähän
4. En lainkaan
5. En osaa sanoa

SC 25. Missä määrin arvelet, että muilla yhteisösi jäsenillä tai muilla kansalaisilla on samat perusarvot, samaistumisen kohteet ja maailmankuva kuin virastolla tai organisaatiolla, jossa työskentelet?
1. Paljon
2. Joskus
3. Vähän
4. En lainkaan
5. En osaa sanoa

SC 26. Missä määrin arvelet, että muilla yhteisösi jäsenillä tai muilla kansalaisilla on samat perusarvot, samaistumisen kohteet ja maailmankuva kuin sinulla suhteessa siihen, mitä maassasi tapahtuu tällä hetkellä JA miten yhteiskunnallisia ongelmia yleensä käsitellään?
1. Paljon
2. Joskus
3. Vähän
4. En lainkaan
5. En osaa sanoa

SC 27. Kuinka *varma* olet siitä, että useimmat ihmiset ymmärtävät, mitä maassasi tällä hetkellä tapahtuu ja mitä tulevaisuus tuo tullessaan sen kansalaisille?
1. Erittäin
2. Jonkin verran
3. Hieman

4. En lainkaan
5. Erittäin epätodennäköistä
6. En osaa sanoa

Teknologia / ICT

T 28. Vuonna 2011 noin 89,7 % suomalaisista käytti internetiä. Kuinka usein itse käytät internetiä hakeaksesi tietoa hallituksesta tai sen toimijoista?
1. Aina 2. Joskus 3. En kovin usein
4. Silloin tällöin 5. En lainkaan

T 29. Kuinka usein haet opetusministeriötä koskevaa tietoa internetistä?
1. Aina 2. Joskus 3. En kovin usein
4. Silloin tällöin 5. En lainkaan

T 30. Kuinka usein haet terveysministeriötä koskevaa tietoa internetistä?
1. Aina 2. Joskus 3. En kovin usein
4. Silloin tällöin 5. En lainkaan

T 31. Mistä saat eniten? tietoa maan politiikasta ja päätöksenteosta?
1. Sanomalehdet 2. Internet 3. Verkkomediat
4. Kännykkä 5. Sähköposti 6. Muu, mikä?

T 32. Mitä seuraavista medioista käytät useimmiten päivittäiseen tai muuhun säännölliseen viestintään?
1. Sähköposti 2. Kännykkä 3. Tekstiviestit
4. Skype 5. Henkilökohtaiset tapaamiset
6. Muu, mikä?

T 33. Mitä seuraavista medioista käytät mieluiten työtovereiden ja/tai kansalaisten kanssa viestimiseen?
1. Sähköposti 2. Kännykkä 3. Tekstiviestit
4. Skype 5. Henkilökohtaiset tapaamiset
6. Muu, mikä?

T 34. Oletko samaa vai eri mieltä siitä, että sähköpostin, matkapuhelinten, tekstiviestien, Skypen ja muiden sähköisten medioiden ja viestimien avulla viestiminen tekee suomalaisesta kansallisen tason päätöksenteosta ja yhteiskunnasta toimivamman?
1. Vahvasti samaa mieltä
2. Samaa mieltä
3. En samaa enkä eri mieltä
4. Eri mieltä
5. Vahvasti eri mieltä
6. En osaa sanoa

T 35. Noin 85 % suomalaista kotitalouksista ja kansalaisista käyttää tietokonetta. Mihin tarkoitukseen käytät itse tietokonetta eniten?
1. Työhön 2. Nettiostosten tekemiseen
3. Pankkiasioiden hoitamiseen 4. Laskujen maksamiseen
5. Yhteydenpitoon 6. Kaikkiin edellisiin

Institutionaalinen tehokkuus / tulokset

IE/O 36. Kuinka usein käytät internetiä suomalaisen koulutuspolitiikan tulosten tai tehokkuuden arvioimiseen?
1. Aina 2. Joskus 3. En kovin usein
4. Silloin tällöin 5. En lainkaan

IE/O 37. Edelliseen kysymykseen (IE/O 35) viitaten, millainen arviosi on yleisesti ottaen ollut?
1. Positiivinen
2. Melko positiivinen
3. Keskiverto
4. Melko negatiivinen
5. Negatiivinen

IE/O 38. Miten arvioisit Suomen koulutuspolitiikan toimivuutta asteikolla 1–5 (jossa 5 on korkein)?
1. Heikko 2. Tyydyttävä 3. Keskiverto
4. Erinomainen 5. Loistava

IE/O 39. Kuinka usein käytät internetiä suomalaisen terveyspolitiikan tulosten tai toimivuuden arvioimiseen?
1. Aina 2. Joskus 3. En kovin usein
4. Silloin tällöin 5. En lainkaan

IE/O 40. Edelliseen kysymykseen (IE/O 38) viitaten, millainen arviosi on yleisesti ottaen ollut?
1. Positiivinen
2. Melko positiivinen
3. Keskiverto
4. Melko negatiivinen
5. Negatiivinen

IE/O 41. Miten arvioisit Suomen terveyspolitiikan toimivuutta asteikolla 1–5 (jossa 5 on korkein)?
1. Heikko 2. Tyydyttävä 3. Keskiverto
4. Erinomainen 5. Loistava

Poliittinen dissonanssi

PD 42. Maan hallituksen kansallisista asioista antama tieto on johdonmukaista ja helposti ymmärrettävää.

1 Vahvasti samaa mieltä
2 Samaa mieltä
3 En samaa enkä eri mieltä
4 Eri mieltä
5 Vahvasti eri mieltä
6 En osaa sanoa

PD 43. Paikallishallinnon paikallisista päätöksentekoasioista antama tieto on johdonmukaista ja helposti ymmärrettävää.

1 Vahvasti samaa mieltä
2 Samaa mieltä
3 En samaa enkä eri mieltä
4 Eri mieltä
5 Vahvasti eri mieltä
6 En osaa sanoa

PD 44. Uskon, että hallituksen ja paikallishallinnon yhteisten prosessien ja yhteisen termistön puute on tällä hetkellä tehokkaan hallinnollisen koordinaation este.

1 Vahvasti samaa mieltä
2 Samaa mieltä
3 En samaa enkä eri mieltä
4 Eri mieltä
5 Vahvasti eri mieltä
6 En osaa sanoa

PD 45. Kuinka usein oman kokemuksesi perusteella käy niin, että paikalliset olosuhteet määrittävät kansallisen tason toimenpiteiden toteutusta tavalla, joka poikkeaa kansallisen tason viranomaisten antamista tarkoista ohjeista?

1 Melkein aina
2 Useimmiten
3 Toisinaan
4 Harvoin
5 En osaa sanoa

PD 46. Mikä on mielestäsi suurin tämänhetkinen tai mahdollinen uhka maallesi?

1. Lama	2. Työttömyys	3. Terrorismi
4. Korruptio	5. Inflaatio	6. Väestönmuutos

PD 47. Mitä seuraavista maan hallituksen pitäisi mielestäsi priorisoida infrastruktuurin kehittämisessä?
1. Vesihuolto
2. Sähköverkko
3. Oppilaitosten uudelleenjärjestely
4. Julkishallinnon vastuullisuuden palauttaminen
5. Hyvä terveydenhuol to
6. Teknologiateollisuuden auttaminen / vahvistaminen
7. Tie- ja kuljetusverkoston uudelleenrakentaminen

PD 48. Mitä seuraavista paikallishallinnon pitäisi mielestäsi priorisoida infrastruktuurin kehittämisessä?
1. Vesihuolto
2. Sähköverkko
3. Oppilaitosten uudelleenjärjestely
4. Julkishallinnon vastuullisuuden palauttaminen
5. Hyvä terveydenhuol to
6. Teknologiateollisuuden auttaminen / vahvistaminen
7. Tie- ja kuljetusverkoston uudelleenrakentaminen

Tietoteoreettinen relativismi

ER 49. Mikä on mielestäsi merkittävin ongelma maailmassa?

Nälkä	Globalisaatio	Ilmastonmuutos
1	2	3
Maahanmuutto	Sodat ja konfliktit	Huume- tai ihmiskauppa
4	5	6

ER 50. Mikä on mielestäsi merkittävin ongelma Suomessa?

Teknologia	Globalisaatio	Ilmastonmuutos
1	2	3
Maahanmuutto	Kulttuurinen muutos	Huume- tai ihmiskauppa
4	5	6

ER 51. Onko meillä mielestäsi tällä hetkellä hyvä kuva ihmiskuntaa kohtaavista lukuisista ongelmista?

Vahvasti eri mieltä	Eri mieltä	En eri mieltä eikä samaa mieltä
1	2	3
Samaa mieltä	Vahvasti samaa mieltä	
4	5	

ER 52. Jos tunnistamme maailmaa kohtaavia ongelmia, uskotko, että meillä on yhteinen kanta tai yhteisiä vastauksia niiden ratkaisemiseen?

Vahvasti eri mieltä	Eri mieltä	En eri mieltä eikä samaa mieltä
1	2	3
Samaa mieltä	Vahvasti samaa mieltä	
4	5	

ER 53. Pitäisikö mielestäsi perinteistä suomalaista kulttuuria ja perinteisiä suomalaisia normeja suojella/säilyttää, kun teknologia ja globalisaatio lisääntyvät?

Vahvasti eri mieltä	Eri mieltä	En eri mieltä eikä samaa mieltä
1	2	3
Samaa mieltä	Vahvasti samaa mieltä	
4	5	

ER 54. Pitäisikö mielestäsi teknologiaa käyttää apuna, jotta suomalainen yhteiskunta sopeutuisi tulevaisuuteen?

Vahvasti eri mieltä	Eri mieltä	En eri mieltä eikä samaa mieltä
1	2	3
Samaa mieltä	Vahvasti samaa mieltä	
4	5	

ER 55. Uskotko, että yhteiskunnallisiin ongelmiin on vielä löytämättä vastauksia ja ratkaisuja ja pitäisikö meidän jatkaa niiden etsimistä?

Vahvasti eri mieltä	Eri mieltä	En eri mieltä eikä samaa mieltä
1	2	3
Samaa mieltä	Vahvasti samaa mieltä	
4	5	

ER 56. Onko totuus maailmasta jo löydetty?

Vahvasti eri mieltä	Eri mieltä	En eri mieltä eikä samaa mieltä
1	2	3
Samaa mieltä	Vahvasti samaa mieltä	
4	5	

ER 57. Teknologia voi auttaa ihmiskuntaa ymmärtämään ympäröivää maailmaa paremmin.

Vahvasti eri mieltä	Eri mieltä	En eri mieltä eikä samaa mieltä
1	2	3
Samaa mieltä	Vahvasti samaa mieltä	
4	5	

ER 58. Teknologia voi tehdä ympäröivästä maailmasta monimutkaisemman ja siksi vaikeamman ymmärtää ja hallita

Vahvasti eri mieltä	Eri mieltä	En eri mieltä eikä samaa mieltä
1	2	3
Samaa mieltä	Vahvasti samaa mieltä	
4	5	

ER 59. Totuus ympäröivästä maailmasta on mikä on, eikä ihminen tai teknologia voi sitä täysin käsittää.

Vahvasti eri mieltä	Eri mieltä	En eri mieltä eikä samaa mieltä
1	2	3
Samaa mieltä	Vahvasti samaa mieltä	
4	5	

ER 60. Muiden etnisten ryhmien arvot ja perinteet ovat yhtä hyviä ja hyväksyttäviä kuin oman etnisen ryhmäni arvot ja perinteet.

Vahvasti eri mieltä	Eri mieltä	En eri mieltä eikä samaa mieltä
1	2	3
Samaa mieltä	Vahvasti samaa mieltä	
4	5	

ER 61. Teknologian potentiaalinen hyöty ihmiskunnalle on paljon suurempi kuin sen potentiaaliset haitat.

Vahvasti eri mieltä	Eri mieltä	En eri mieltä eikä samaa mieltä
1	2	3
Samaa mieltä	Vahvasti samaa mieltä	
4	5	

Kommentteja: [Tähän voit kirjoittaa vapaasti omia kommenttejasi]

Appendix D

The Parliament Building of Finland (The Eduskunta).

Appendix E

“Lion of Finland without the Crown.” An allegorical sculpture by Gunnar Finne (1886–1952) on the wall of the Parliament Cafeteria.

Appendix F

Author Alighting the Steps in Front of the Parliament of Finland.

Appendix G

Inside the Plenary Hall (Legislative Chamber) of the Parliament of Finland.

Appendix H

Author with the Chairwoman of the Grand Committee of the Parliament of Finland (far right) and Terhi Mölsa (middle), Executive Director of the Fulbright Center, Finland.

Appendix I

“Work and the Future” sculpture by Wäinö Aaltonen (1894–1966) at the Plenary Hall of the Parliament of Finland.

Appendix J

Author on a Ferry Crossing the Gulf of Finland en Route from Helsinki to Tallin (Republic of Estonia).

Appendix K

Author Astride the Arctic Circle Line in the Northern Laplands, Finland,
Latitude 66° 32' 35" N of the Equator.

Appendix L

Author Facing the Arctic Circle Line, Carrying Survey Questionnaires.

Appendix M

Questionnaire/Survey Templates Displayed at the Main Public Library, Tampere, Finland.

Appendix N

“Finnish True Grit” Sculpture of a Handyman Near the Bridge across the Kemi River, Rovaniemi, Northern Laplands, Finland. A symbolic personification of the Finnish attitude toward the toil of work, endurance, and the triumph of man versus the natural challenges of his environment.

Appendix O

Author in front of the Arctic Circle Signpost, 2013.

Bibliography

Ala-Rämi, K. 2007. Communication and distance in collaboration between high technology enterprises in Northern Finland. *European Planning Studies,* 15(8): 1047–1062.

Argyris, C. and D. Schon. 1996. *Organizational learning II: Theory, method and Practice*. Reading, MA: Addison Wesley.

Asplund, R. 2003. *Flexibility and competitiveness: Labour market flexibility, innovation, and organizational performance*. Helsinki, Finland: Taloustieto Oy.

Baines, L. 2007. Learning from the world: Achieving more by doing less. *Phi Delta Kappan*, 89(2): 98–100.

Beebe T. J., P. A. Harrison, A. Sharma, and S. Hedger. 2001. The community readiness survey: Development and initial validation. *Evaluation Review,* 25(1): 55–71.

Bosworth, M. 2013. Why Finland loves saunas. *Magazine*, October 1.

Cheng, W-P. 2001. *A strategic approach of building national technology strength: Finland as case study.* Master Thesis, Sun Yat-sen University, Taiwan, p. 16. Also see CRN: etd-0627101–121030.

Clift, B. 2006. The new political economy of dirigisme: French macroeconomic policy, unrepentant sinning and the stability and growth pact. *British Journal of Politics and International Relations,* 8 (3): 388–409.

Coleman, J. S. 1990. *Foundations of social theory*. Cambridge, UK: Belknap Press.

Colvin, G. 1999. We see the future, and it's Finland. *Fortune,* 24(4): 1–4.

Committee for the Future, The. 2013. *Black swans: What will change the world next?* Helsinki, Finland: The Finnish Parliament's Committee for the Future.

Committee for the Future, The. 2015. 100 Opportunities for Finland and the world: Radical technology inquirer (RTI) for anticipation/evaluation of technological breakthroughs. Helsinki, Finland: Committee for the Future, The Parliament of Finland.

Cronbach L. 1951. Coefficient alpha and the internal structure of tests. *Psychometrika,* 44: 373–393.

Dahl, R. 1985. *Controlling nuclear weapons: Democracy versus guardianship.* Syracuse, NY: Syracuse University Press.

Dator, J. 2007. Governing the futures: Dream or survival societies? In *Human competency development in the information society*, ed. J. Kasvi, P. Tiihonen, and K. Nolvi, 15–24. Helsinki Finland: Proceedings of the Fifth General Assembly of the International Parliamentarians' Association for Information Technology (IPAIT V), Committee for the Future, Parliament of Finland.

Daveri, F. and O. Silva. 2004. Not only Nokia: what Finland tells us about the new economy. *Economic Policy* (April): 117–163.

Delbridge, R. and T. Edwards. 2013. Inhabiting institutions: Critical realist refinements to understanding institutional complexity and change. *Organization Studies*, 34(7): 927–947.

Dickey, C. 2013. Iranian bombs and black swans. *The Daily Beast*, December 17, 2013, p. 1. http://news.yahoo.com/iranian-bombs-black-swans-104500828—politics.html. Accessed on August 10, 2015.

Doupi, P, E. Renko, P. Hämäläinen, M. Mäkelä, S. Geist, and J. Dumortier. 2010. E-health strategies country brief: Finland. Brussels, Belgium: European Commission.

Eagly A. H. 1978. Sex differences in influencability. *Psychological Bulletin*, 85: 86–116.

Economic Intelligence Unit. 2014. The learning curve: Education and skills for life 2014 report, New York, NY: Pearson.

Economist, The. 2014. What China wants, August 23, 2014.

Economist, The. 2014. The People's Bank of China, November 22, 2014.

Economist, The. 2014. Russia's wounded economy, November 22, 2014.

Economist, The. 2015. An hereditary meritocracy. January 24, 2015.

Economist, The. 2015. Excellence v equity, Special report on universities, March 28, 2015.

Edgerton, D. 2009. The shock of the old. In *Technology and the future*, ed. A. H. Teich, 63–73. Boston, MA: Wadsworth.

Eduskunta (Parliament of Finland). 2012. The committee for the future's vision, strategy and operational plan for 2011–15. *Draft*, April 17, p. 3.

Elias L. 2008. *The Kalevala* (Translated by Keith Bosley). New York, NY: Oxford University Press.

Emanuel, E. and T. Spiro. 2015. The coming health-care cost shock, *The Wall Street Journal*, July 8.

European Commission. 2001. European competitiveness report 2001. Luxembourg: The Office for Official Publications of the European Communities.

Eurostat, OECD. 2013.

Finnish Academies of Technology. 2004. *High technology Finland* 2004. Helsinki, Finland: Finnish Academies of Technology.

Finnish Communications Regulatory Authority (FICORA). 2011. The revolution in communications: Will all consumers make the switch to the Internet? Helsinki, Finland: FICORA.

Finnish Institute of Occupational Health. 2014. The status of occupational health services in Finland and the role of the Finnish Institute of Occupational Health in the development of occupational health services: Report for the international evaluation of the Finnish Institute of Occupational Health (FIOH).

Frank, R. H. and B. S. Bernanke. 2007. *Principles of macroeconomics*, 3rd ed. Boston, MA: McGraw-Hill.

Graig, L. A. 1999. *Health of nations: An international perspective on U.S. health care reform*. Washington, DC: CQ Press.

Granato, J., R. Inglehart, and D. Lange. 1996. The effects of cultural values on economic development: Theory, hypotheses, and some empirical tests. *American Journal of Political Science*, 40(3): 607–631.

Gregg, P. M. and A. S. Banks. 1965. Dimensions of political systems: Factor analysis of a cross-polity survey. *American Political Science Review*, 59: 602–614.

Häkkinen, U. 2005. The impact of changes in Finland's health care system. *Health Economics*, 14: 101–118.

Häkkinen, U. 2010. Financing of hospital care in Finland. *Euro Observer*, 12(3): 10–12.

Hamalainen, T. 2013. *Building blocks for a new socioeconomic model*. Helsinki, Finland: SITRA, Version 10.

Hancock, L. 2011. Why are Finland's schools successful? *Smithsonian Magazine*, September.

Haythornthwaite, C. 2005. Social networks and Internet connectivity effects. *Information. Communications & Society,* 8: 125–147.

Hertzman, C, J. Frank, and R. C. Evans. 1994. Heterogenetics in Health Status and the determinants of population health. In *Why are some people healthy and others not*? ed. R. G. Evans, M. L. Barer, and T. R. Marmor, 67–92. New York, NY: Aldine De Gruyter.

Herring, P. 2004. High technology Finland 2004. Helsinki, Finland: Finnish Academies of Technology.

Herring, P. 2003. Inspired by ideas. *High technology Finland* 2003. Helsinki, Finland: High Technology Finland.

Herring, P. 2013. *HighTech Finland* 2013. Helsinki, Finland: HighTech Finland.

HighTech Finland. 2008. *A world of ideas and innovation.* Helsinki, Finland: HighTech Finland.

High Technology Finland. 2003. Communications and information. *High technology Finland.* Helsinki, Finland: High Technology Finland.

Hilsenrath, J. and B. Blackstone. 2015. Cheaper currency has global impact. *The Wall Street Journal*, August 12.

Hofstede, G. 1984. *Culture's consequences*. Newbury Park, CA: Sage Publications.

Hofstede, G. 1999. Problems remain, but theories will change: The universal and the specific in 21st –century global management. *Organizational Dynamics,* 28(1): 34–44.

Holstila, E. 2004. High technology Finland: Creative people are the critical success factor. Helsinki, Finland: Finnish Academies of Technology.

Ilmavirta, V. 2004. High technology Finland 2004: An essential link in Finland's innovation chain. Helsinki, Finland: Finnish Academies of Technology.

Inglehart, R. and P. R. Abramson. 1994. Economic security and value change. *American Political Science Review,* 88(2): 336–354.

Inglehart, R. and W. E. Baker. 2000. Modernization, cultural change, and the persistence of traditional values. *American Sociological Review,* 65: 19–51.

Inkinen, T. 2006. The social construction of the urban use of information technology: The case of Tampere, Finland. *Journal of Urban Technology,* 13(3): 49–75.

Jakubowski, E. and R. Busse. 1998. Health care system in the E U: A comparative study. Luxembourg: Directorate General for Research, European Parliament.

Jaruzelski. Statement credited to Barry Jaruzelski, Senior Partner at Booz & Company. http://www.wipo.int/pressroom/en/articles/2013/article_0016.html. Accessed on August 10, 2015.

Järvelin, J. 2002. Health care system in transition. *European Observatory on Health Care Systems.*

Järvelin, J. 2006. Far-reaching decisions needed in a time of transition. *Finland decide to succeed*! Helsinki, Finland: Secretariat of the Committee for the Future, The duskunta. Jauhiainen, J. S. 2006. Multipolis: High technology networks in northern Finland. *European Planning Studies,* 14(10): 1407–1428.

Kalu, K. N. 2005. Competing ideals and the public agenda in Medicare reform: The 'garbage can' model revisited, *Administration & Society,* 37(1): 23–56.

Kalu, K. N. 2011. Institution-building, not nation-building: a structural-functional model. *International Review of Administrative Sciences,* 77(1): 119–137.

Kalu N. K. and B. Remkus. 2010. The evolution of social capital and civic engagement between nonprofit networks and county representatives: A social constructivist approach. *Social Science Computer Review,* 28(1): 135–150.

Karjula, K. 2006. To the reader: *Finland, decide to succeed! A background polemic on innovation policy*, Helsinki, Finland: Secretariat of the Committee for the Future, The Eduskunta.

Kautonen, M. and J. Kolehmainen. 2001. Perspectives on regional innovation policy of the learning economy. In *Alueiden kilpailukyvyn kahdeksan elementtiä*, ed. M. Sotarauta, and N. Mustikkamäki, 109–128. Helsinki, Finland: Suomen Kuntaliitto.

Kavanaugh, A. 2001. Civic engagement among community computer network users: Trend or phrase? Research conference on communication, information, and Internet policy. Alexandria, VA: Telecommunications Policy Research Conference.

Keesing. R. M. 1974. Theories of culture. *Annual Review of Anthropology*, (3): 73–97.

Kelly, G. 2013. Finland and Nokia: An affair to remember. *Business*, October 4, 2013. http://www.wired.co.uk/news/archive/2013–10/04/finland-and-nokia.

Kim, J. O. and C. W. Mueller. 1978. *Introduction to factor analysis: What it is and how to do it*. Newbury Park, CA: Sage.

Korhonen-Yrjanheikki, K. 2007. New challenging approaches to engineering education: Enhancing university-industry co-operation. *European Journal of Engineering Education*, 32(2): 167–179.

Kuittinen, T. 2013. A brief history of Nokia: When the future was Finnish. http:brg.com/2013/09/03/nokia-history-analysis-microsoft-acquisition,September 3, 2013, p. 2. Accessed on June 15, 2015.

Lamb, R. 2009. What would nuclear winter be like? 13 January 2009.

Leidner, D. E. and T. Kayworth. 2006. A review of culture in information systems research: Toward a theory of information technology culture conflict. *MIS Quarterly*, 30(2): 357–399.

Lester, R. K. and M. Sotarauta. 2007. Executive summary: Innovation, universities, and the competitiveness of regions, *Technology review, 214*. Helsinki, Finland: Tekes.

Lipponen, P. 2012. The Committee of the Future is one of a kind. Paper presented at The 28th IFLA Pre-Conference of Library and Research Services for Parliaments, Helsinki, Finland, August 9, 2012, p. 4. (Päivi Lipponen is a member of the Finnish Parliament and Chair of the Committee for the Future).

Lundvall, B-A., ed. 1992. *National systems of innovation: Toward a theory of innovation and interactive learning*. London, UK: Pinter.

Lupton, D. 1995. *The imperative of health: Public health and the regulated body*. Thousand Oaks, CA: Sage.

Mann, H. B. and D. R. Whitney. 1947. One a test of whether one of two random variables is stochastically larger than the other. *Annals of Mathematical Statistics*, 18: 50–60.

Marten, P., ed. 2012. *Welcome to Finland: The design nation*, Helsinki, Finland: The Ministry of Foreign Affairs.

McArdle, M. 2013. As goes Nokia, so goes Finland? *Bloomberg View*, September 3, 2013, p. 1–2. http://www.bloombergview.com/articles/2013/-09-03/as-goes-nokia-so-goes-finland. Accessed on June 15, 2015.

Mesthene, E. G. 2009. The role of technology in society. In *Technology and the future*, ed. A. H. Teich, 77–86. Boston, MA: Wadsworth.

Ministry of Education and Culture. 2013. *The Finnish education and decision making system*. Helsinki, Finland.

Ministry of Social Affairs and Health. 2010. Toward social sustainability, *Social sustainable Finland 2020: Strategy for social and health policy*, December.

Ministry of Social Affairs and Health. 2011. Challenges and opportunities of the 2010s. Helsinki, Finland.

Moen, E. and K. Lilja. 2005. Change in coordinated market economies: The case of Nokia and Finland. In *Changing capitalisms: Internationalization, institutional change and systems of economic organization*, ed. G. Morgan, R. Whitley and E. Moen, 352–379. Oxford: Oxford University Press.

Morgan, G. 2006. *Images of organization*. Thousand Oaks, CA: Sage Publications.

Mukkala, K. 2010. The role of regional policies in promoting networking and innovative activity: Evidence from small Finnish high-tech firms. *European Planning Studies*, 18(7): 1059.

Murto, E., M. Niemela, and T. Laamanen. 2006. *Finnish technology policy from the 1960s to the present day: Public sector activities promoting research and development*. Helsinki, Finland: Ministry of Trade and Industry.

Nunnally, J. and H. Bernstein. 1994. *Psychometric theory*, 3rd ed. New York, NY: McGraw-Hill.

Nykanen, P. and H. Paulapuro. 2005. *Around the roll: A century of Finnish paper machine and papermaking technology*. Helsinki, Finland: The Society for History of Technology.

OECD. 2013. Equality and quality: Ensuring educational opportunity for all. *Education policy outlook: Finland* www.oecd.org/policyoutlook.htm.

OECD. 2013. Governance: Autonomous decision making at the local level. *Education policy outlook: Finland* www.oecd.org/policyoutlook.htm.

OECD. 2013. Preparing students for the future: Well-developed general and vocational education and training programmes, *Education policy outlook: Finland* www.oecd.org/policyoutlook.htm.

OECD. 2013. School improvement: Trust and proficiency of school leaders and teachers. *Education policy outlook: Finland* www.oecd.org/policyoutlook.htm.

Ornston, D. 2006. Reorganizing adjustment: Finland's emergence as high tech leader. *West European Politics*, 29(4): 784–801.

Ovide, S. 2015. Microsoft's misdial cost $7.6 billion. *The Wall Street Journal*, Vol. CCLXVI (7), Thursday July 9, 2015.

Parliament of Finland. 2000. *The voice of the people: Past, present, and future*. Helsinki, Finland: Parliament of Finland.

Parliament of Finland. 2011. *Parliament's buildings and art*. Helsinki: The Parliament of Finland.

Parliament of Finland. 2012. The Committee for the Future's vision, strategy and operational plan for 2011–15. *Draft*, April 17.

Parliament of Finland. 2013. *Democracy's long road*. Helsinki, Finland: The Eduskunta.

Parsons, T. 1937. *The structure of social action*. New York, NY: McGraw Hill.

Parsons, T. 1956. A sociological approach to the theory of organizations. In *Structure and processes in modern societies*, ed. Talcott Parsons, 27. Glencoe, IL: Free Press.

Patrick, D. L. and T. M. Wickizer. 1995. Community health. In *Society and health*, ed. B. C. Amick III, S. Levine, A. R. Tarlov, and D. C. Walsh, 46–92. New York, NY: Oxford University Press.

Patrucco, P. P. 2003. Institutional variety, networking and knowledge exchange: Communication and innovation in the case of Brianza technological district. *Regional Studies*, 37(2): 159–172.

Perkiö-Mäkelä et al. 2006. Work and health survey 2006, *Table report*. Tampere, Finland: Finnish Institute of Occupational Health.

Pollock, P. H. 2012. *The essentials of political analysis*. New York, NY: CQ Press, 164.

Powell, N. 2011. How Finland's economy became hooked on Nokia. *The Globe and Mail*, October 26, 2011.

Robert, P. 1993. *Making democracy work: Civic traditions in modern Italy*. Princeton, NJ: Princeton University Press.

Robert P. 1995. Bowling alone: America's declining social capital. *Journal of Democracy,* 6: 65–78.

Rawls, J. 1992. Justice as fairness: Political not metaphysical. In *Communitarianism and individualism*, ed. S. Avineri and A. de-Shalit, 186–204. New York, NY: Oxford University Press.

Room, G. 2011. *Complexity, institutions, and public policy: Agile decision-making in a turbulent world.* Cheltenham, UK: Edward Elgar.

Rouvinen, P. and P. Yla-Antilla. 2003. Little Finland's transformation to a wireless giant. In *The global information technology report* 2003–2004, ed. S. Dutta, B. Lanvin, and F. Paua, 87–109. New York, NY: Oxford University Press.

Rummel, R. J. 1967. Understanding factor analysis. *The Journal of Conflict Resolution,* 11: 444–480.

Ruohonen, A., S. Salzano, and P. Marten. 2014. Bring on the Finnish summer festivals. *This is Finland*, pp. 1–6. http://finland.fi/Public/default.aspx?contentid=307471&nodeid=44497&culture=en-US. Accessed on August 12, 2015.

Saarivirta, T., D. Consoli, and P. Dhondt. 2012. The evolution of the Finnish health-care system early 19th century and onwards. *International Journal of Business and Social Science,* 36(6): 243–257.

Sahlberg, P. 2011. *Finnish lessons: What can the world learn from educational change in Finland?* New York, NY: Columbia University.

Schienstock, G. and T. Hamalainen. 2001. *Transformation of the Finnish innovation system: A network approach.* Helsinki, Finland: Finnish National Fund for Research and Development.

Sen, A. 2006. *Identity and violence: The illusion of destiny*. New York, NY: W. W. Norton & Co.

Simnett, I. 1995. *Managing health promotion: Developing healthy organizations and communities*. New York, NY: John Wiley & Sons.

Sipilä, K. 2003. Inventions create a better future. *High technology Finland* 2003. Helsinki, Finland: High Technology Finland.

Smil, V. 2008. *Global catastrophes and trends: The next 50 years*. Cambridge, MA: MIT Press.

Sorakunas, E. 2004. *HighTechnology Finland.* 2004: Rooted in history, looking to the future. Helsinki, Finland: Finnish Academies of Technology.

Statistics of Finland. 2013. Bureau of Economic Analysis, September 30.

Stenbock, D. 2000. *The Nokia revolution: The story of an extraordinary company that transformed an industry*. New York, NY: Amacom.

Stig, K., I. P. Lütz, and the Nordic Reference Group. 2013. *Financing health care in the Nordic countries*. Copenhagen, Denmark: Nordic Medico Statistical Committee.

Stoessinger, J. G. 2010. *Why nations go to war*. Belmont, CA: Wadsworth/Thomson Learning.

Suurla, R., M. Markkula, O. Mustajarvi. 2002. *Developing and implementing knowledge management in the parliament of Finland.* Helsinki, Finland: Parliament of Finland, Committee for the Future (Picture 41), 147.

Suvikumpu, L. 2010. *Find out about Finland.* Keuruu, Finland: Otava Pu*blis*hing Co.Limited.

Tahvannainen, A-J. 2011. Diagnosing the growing pains of a technology-based industry: An examination of the Finnish biotechnology industry in light of empirical economics. Helsinki, Finland: The Research Institute of the Finnish Economy, 18–40.

Taleb, N. N. 2007. *The black swan: The impact of the highly improbable*. New York, NY: Random House, Inc.

Taylor, A. 2012. Why Finland's unorthodox education system is the best in the world. *Business Insider*. http://www.businessinsider.com/finlands-education-system-best-in-world-2012-11?op=1#ixzz3aADmiNAq.

Teperi, J., M. E. Porter, L. Vuorenkoski, and J. F. Baron. 2009. The Finnish health care system: A value-based perspective. *SITRA Reports 82*. Helsinki, Finland.

Tiihonen, P. 2011. *Revamping the work of the committee for the future: Summary*. Helsinki, Finland: Committee for the Future, Parliament of Finland.

Tiihonen, P. 2014. A committee for the future. *Future Justice*, February 17. www.futurejustice.org/blog/guest-contribution/guest-article-a-committee-for-the-future. Accessed on July 24, 2105.

Tiihonen, S. 2008. The public sector in the conflicting pressures of democracy and efficiency. In *Very many democracies*, ed. P. Tiihonen, 53–85. Helsinki, Finland: Committee of the Future, Parliament of Finland, 66 (April).

Trice, H. M. and J. M. Beyer. 1993. *The culture of work organizations*. Upper Saddle River, NJ: Prentice-Hall.

Venkatesh V., M. G. Morris, P. Ackerman, and T. Sykes. 2004. Individual reactions to new technologies in the workplace: The role of gender as a psychological construct. *Journal of Applied Social Psychology*, 34(3): 445–467.

Voegelin, E. 1957. *Order and history: Plato and Aristotle*. Baton Rouge, LA: Louisiana State University.

Vuorenkoski, L., P. Mladovsky, and E. Mossialos. 2008. Finland: Health systems review, *Health Systems in Transition*, 10(4): 1–168.

Wiki Countries Navigation. 2015.

World Health Organization. 2012. National profile of occupational health system in Finland. Copenhagen, Denmark: Regional Office for Europe, p. x.

World Intellectual Property Organization (WIPO). 2013. Global innovation index 2013. http://www.wipo.int/pressroom/en/articles/2013/article_0016.html, p. 6.

Wuorinen, J. H. 1965. *A History of Finland*. New York, NY: The American-Scandinavian Foundation.

Digital Sources

Amberla, K. 1999. Facts about Finnish music life, *Finnish Music Quarterly* 3, September, p. 1. http://fmq.fi/1999/09/facts-about-finnish-music-life.

CIA World Factbook. http://www.cia.gov/library/publications/the-world-factbook/geos/fi.html. Accessed on August 8, 2015.

CIMO. 2014. Educational Authorities in Finland. www.studyinfinland.fi/destination_finland/education_system/educational_authorities. Accessed on July 29, 2015.

Encyclopedia Britannica. http://www.britannica.com/topic/dirigisme, p. 1. Accessed on July 3, 2015.

Encyclopedia of the Nations. http://www.nationsencylopedia.com/Europe/Finland-Government.html, 3. Accessed on May 31, 2015.

Encyclopedia of the Nations. Finland – Government. http://www.nationsencylopedia.com/Europe/Finland-Government.html.

The Europe 2020 Strategy can be located at: http://ec.europe.eu/europe2020/pdf/csr2014/nrp2014_finland_en.pdf, Accessed on July 29, 2015.

European Commission, http://ec.europa.eu/programmes/horizon2020/en/h2020-section/future-and-emerging-technologies. Accessed on July 24, 2105.

The Hofstede Center, Dimensions of National Culture. www.hofstede.com.

HowStuffWorks.com. http://science.howstuffworks.com/nuclear-winter.htm. Accessed on August 10, 2015.

http://www.britannica.com/biography/Aleksis-Kivi, p. 1. Accessed on August 12, 2015.

http://www.britannica.com/print/article/207424, p. 4. Accessed on August 12, 2015.

http://www.britannica.com/print/article/207424, p. 6. Accessed on August 12, 2015.

http://www.euro.who.int/__data/assets/pdf_file/0009/113877/E93944.pdf?ua=1, Accessed on August 5, 2015.

http://www.everyculture.com/Cr-GA/Finland.html, p. 7. Accessed on August 12, 2015.

http://www.everyculture.com/Cr-GA/Finland.html, p. 8. Accessed on August 12, 2015.

http://www.fime.fi/en/culture/culture-in-finland. Accessed on August 112, 2015.

http://finland.fi/Public/default.aspx?contentid=16007&nodeid=44495&culture=en-US, p. 2. Accessed on August 12, 2015.

http://finland.fi/Public/default.aspx?contentid=209426&nodeid=44495&culture=en-US, p. 2. Accessed on August 12, 2015.

http://geert-hofstede.com/finland.html. Accessed on November 8, 2015.

https://www.google.com/?gws_rd=ssl#q=unto+Hakkinen%2C+Financing+of+hospital+care+in+Finland, p. 10. Accessed on August 15, 2015.

http://www.history.com/topics/new-deal, pp. 1–7. Accessed on July 3, 2015.

http://independenttravelcats.com/2015/03/22/visiing-finland-in-winter-top-winter-activities-in-finland. Accessed on August 18, 2015.

http://karenrager.tripod.com/essays/kalevala.html. Accessed on August 12, 2015.

http://matadornetwork.com/life/x-things-not-know-finland. Accessed on August 16, 2015.

http://www.med.uottawa.ca/sim/data/Primary_Care.htm. Accessed on August 5, 2015.

http://www.newworldencyclopedia.org/entry/Finland; http://en.wikipedia.org/wiki/Finland.

http://www.photius.com/countries/finland/society/finland_society_income_security_prog_13.html, p. 2. Accessed on August 1, 2015.

http://pureinsideout.com/sauna-finnish.html. Accessed on August 16, 2015.

Kiviniemi, T. O. et al. 2008. Effects of cognac on coronary flow reserve and plasma antioxidant status in healthy young men. *Cardiovascular Ultrasound* 6: 25. http://www.ncbi.nlm.nih.gov/pmc/articles/PMC2427009, 1-9, p. 1.Accessed on July 31, 2015.

Nokia. 2015. Our story. http://company.nokia.com/en/about-us/our-company/our-story, p. 1–2. Accessed on June 15, 2015.

Ojaniemi, T. 2010. Coffee as a Finnish institution. *Finnish Institutions research paper*, FAST Area Studies Program, Department of Translation Studies, University of Tampere, p. 1 (pp. 1–7). http://www15.uta.fi/FAST/FIN/GEN/to-coffee.html. Accessed on July 31, 2015, p. 1.

U.S. State Department, The. 2007. See http://www.infoplease.com/country/profiles/finland.html. Accessed on August 15, 2015.

YouTube. https://www.youtube.com/watch?v=5pQMW8UQYZw&feature=youtube.

Index

For Product Safety Concerns and Information please contact our EU representative GPSR@taylorandfrancis.com
Taylor & Francis Verlag GmbH, Kaufingerstraße 24, 80331 München, Germany

www.ingramcontent.com/pod-product-compliance
Ingram Content Group UK Ltd.
Pitfield, Milton Keynes, MK11 3LW, UK
UKHW022203190926
475702UK00003B/178

* 9 7 8 1 4 8 2 2 2 3 4 4 6 *